T0351959

BIO-INSPIRED
FLYING ROBOTS
EXPERIMENTAL SYNTHESIS OF
AUTONOMOUS INDOOR FLYERS

BIO-INSPIRED FLYING ROBOTS

EXPERIMENTAL SYNTHESIS OF AUTONOMOUS INDOOR FLYERS

Jean-Christophe Zufferey

EPFL Press

A Swiss academic publisher distributed by CRC Press

CRC Press
Taylor & Francis Group

Taylor and Francis Group, LLC
6000 Broken Sound Parkway, NW, Suite 300,
Boca Raton, FL 33487

Distribution and Customer Service
orders@crcpress.com

www.crcpress.com

Library of Congress Cataloging-in-Publication Data
A catalog record for this book is available from the Library of Congress.

This book is published under the editorial direction of Professor Peter Ryser.

For the work described in this book, Dr. Jean-Christophe Zufferey was awarded the "EPFL Press Distinction", an official prize discerned annually at the Ecole polytechnique fédérale de Lausanne (EPFL) and sponsored by the Presses polytechniques et universitaires romandes. The Distinction is given to the author of a doctoral thesis deemed to have outstanding editorial, instructive and scientific qualities; the award consists of the publication of this present book.

is an imprint owned by Presses polytechniques et universitaires romandes, a Swiss academic publishing company whose main purpose is to publish the teaching and research works of the Ecole polytechnique fédérale de Lausanne.

Presses polytechniques et universitaires romandes
EPFL – Centre Midi
Post office box 119
CH-1015 Lausanne, Switzerland
E-Mail: ppur@epfl.ch
Phone: 021 / 693 21 30
Fax: 021 / 693 40 27

www.epflpress.org

© 2008, First edition, EPFL Press
ISBN 2-940222-19-3 (EPFL Press)
ISBN 978-1-4200-6684-5 (CRC Press)

Printed in the UK

All right reserved (including those of translation into other languages). No part of this book may be reproduced in any form – by photoprint, microfilm, or any other means – nor transmitted or translated into a machine language without written permission from the publisher.

Dans la certitude de quels ciels,
au coeur de quels frêles espaces,
étreindre les chauds reflets de vivre ?

Frémis,
Matière qui t'éveilles,
scintille plutôt que ne luis,
tremble comme le milieu de la flamme,
Matière qui virevoltes et t'enfuis
et,
parmi les vents illimités de la conscience,
aspires à être...

Julien Zufferey

Preface

Indoor flying robots represent a largely unexplored area of robotics. There are several unmanned aerial vehicles, but these are machines that require precise information on their absolute position and can fly only in open skies far away from any object. Flying within, or among buildings requires completely different types of sensors and control strategies because geo-position information is no longer available in closed and cluttered environments. At the same time, the small space between obstacles calls for extreme miniaturization and imposes stringent constraints on energetic requirements and mechatronic design.

A small number of scientists and engineers have started to look at flying insects as a source of inspiration for the design of indoor flying robots. But where does one start? Should the robot look like an insect? Is it possible to tackle the problem of perception and control separately from the problem of hardware design? What types of sensors should be used? How do insects translate sensory information in motor commands? These and many other questions are clearly addressed in this book as the author progresses towards the solution of the puzzle.

Biological inspiration is a tricky business. The technology, so to speak, used by biological organisms (deformable tissues, muscles, elastic frameworks, pervasive sensory arrays) differs greatly from that of today's robots, which are mostly made of rigid structures, gears and wheels, and comparatively few sensors. Therefore, what seems effective and efficient in biology may turn out to be fragile, difficult to manufacture, and hard to control in a robot. For example, it is still very debated to which extent robots with rigid legged locomotion are better than robots with articulated wheels.

Also, the morphologies, materials, and brains of biological organisms co-evolve to match the environmental challenges at the spatial and temporal scales where those organisms operate. Isolating a specific biological solution and transposing it into a context that does not match the selection criteria for which that solution was evolved may result in sub-optimal solutions. For example, the single-lens camera with small field of view and high resolution that mammalian brains evolved for shape recognition may not be the most efficient solution for a micro-robot whose sole purpose is to rapidly avoid obstacles on its course.

Useful practice of biological inspiration requires a series of careful steps: (a) describing the challenge faced by robots with established engineering design principles; (b) uniquely identifying the biological functionality that is required by the robot; (c) understanding the biological mechanisms responsible for that functionality; (d) extracting the principles of biological design at a level that abstracts from the technological details; (e) translating those principles into technological developments through standard engineering procedures; and (f) objectively assessing the performance of the robot.

Beside the fascinating results described by the author, this book provides an excellent example of biologically inspired robotics because it clearly documents how the steps mentioned above translate practically into specific choices. This book is also a unique documentary on the entire process of conceiving a robot capable of going where no other robot went before. As one reads through the pages, images of the author come to mind devouring books on flying robots and insects; traveling to visit biologists that culture houseflies and honeybees; spending days in the lab putting together the prototypes and implementing the control circuits; and then finally analyzing the flying abilities of his robots just as his fellow biologists do with insects.

Technology and science will continue to progress, and flying robots will become even smaller and more autonomous in the future. But the ideas, pioneering results, and adventure described in this book will continue to make it a fascinating reading for many years to come.

Dario Floreano

Foreword

The work presented in this book is largely derived from my thesis project, funded by the Swiss National Foundation and carried out at the Swiss Federal Institute of Technology in Lausanne (EPFL), in the Laboratory of Intelligent Systems (http://lis.epfl.ch), under the supervision of Prof. Dario Floreano. This has been a great time during which I had the opportunity to conjugate two of my passions: aviation and robotics. As an aerobatic and mountain-landing pilot, I often felt challenged by these small insects that buzz around flawlessly while exploring their highly cluttered environment and suddenly decide to land on an improbable protuberance. We, as humans, need charts, reconnaissance, weather forecasts, navigational aids; whereas they, as insects, just need the wish to fly and land, and can do it with a brain that has one million times fewer neurons than ours. This is at the same time highly frustrating and motivating: frustrating because engineers have been unable to reproduce artificial systems that can display the tenth of the agility of a fly; motivating because it means that if insects can do it with such a low number of neurons, a way must exist of doing it simply and small. This is why I have been compelled towards a better understanding of the internal functioning of flying insects in order to extract principles that can help synthesize autonomous artificial flyers. Of course it has not been possible to reach the level of expertise and agility of an actual fly within these few years of research, but I hope that this book humbly contributes to this endeavor by relating my hands-on experiments and results. Since (moving) images are better than thousands of (static) words, especially when it comes to mobile robotics, I decided to create and maintain a webpage at http://book.zuff.info containing a list of links, software and videos related to artificial, most of the time bio-inspired, flying robots. I hope it will help you feeling the magic atmosphere surrounding the creation of autonomous flyers.

Of course, I did not spend these years of research completely alone. Many colleagues, undergraduate students and friends have contributed to the adventure and I am sorry not being able to name all them here. However, I would like to cite a few, such as Jean-Daniel Nicoud, André Guignard, Cyril Halter and Adam Klaptocz who helped me enormously with the construction of the microflyers; Antoine Beyeler and Claudio Mattiussi with whom I had countless discussions on the scientific aspects; Adam and Antoine, along with Markus Waibel and Céline Ray, also contributed with many constructive comments on the early manuscripts of this text. External advisors and renowned professors, especially Roland Siegwart, Mandyam Srinivasan, and Nicolas Franceschini, have been key motivators in the fields of mobile and bio-inspired robotics. I also would like to express my gratitude to my parents and family for their patience, for nurturing my intellectual interests since the very beginning, and for their ongoing critical insight. Finally, I would like to thank Céline for her love, support, and understanding, and for making everything worthwhile.

Contents

Contents

CHAPTER 1

Introduction

Flies are objectionable in many ways, but they now add insult to injury by showing that it is definitely possible to achieve the smartest sensory-motor behavior such as 3D navigation at 500 body-lengths per second using quite modest processing resources.

N. Franceschini, 2004

1.1 What's Wrong with Flying Robots?

Current instances of unmanned aerial vehicles (UAV) tend to fly far away from any obstacles, such as ground, trees, and buildings. This is mainly due to aerial platforms featuring such tremendous constraints in terms of manoeuvrability and weight that enabling them to actively avoid collisions in cluttered or confined environments is highly challenging. Very often, researchers and developers use GPS (Global Positioning System) as the main source of sensing information to achieve what is commonly known as "way-point navigation". By carefully choosing the way-points in advance, it is easy to make sure that the resulting path will be free of static obstacles. It is indeed striking to see how research in flying robotics has evolved since the availability of GPS during the mid-1990's[1]. GPS enables a flying robot to

---

[1] After four years of competition, the first autonomous completion of an object retrieval task at the International Aerial Robotics Competition occurred in 1995 and was performed by the Standford team who was the first to use a (differential) GPS.

be aware of its state with respect to a global inertial coordinate system and — in some respects — to be considered as an end-effector of a robotic arm that has a certain workspace in which it can be precisely positioned. Although localisation and obstacle avoidance are two central themes in terrestrial robotics research, they have been somewhat ignored in the aerial robotics community, since it was possible to effortlessly solve the first one by the use of GPS and ignore the second as the sky is far less obstructed than the Earth surface.

However, GPS has several limitations when it comes to low-altitude or indoor flight. The signal sent by the satellites may indeed become too weak, be temporary occluded, or suffer from multiple reflections when reaching the receiver. It is therefore generally admitted that GPS is unreliable when flying in urban canyons, under trees or within buildings. In these situations, the problem of controlling a flying robot becomes very delicate. Some researchers use ground-based beacons or tracking systems to replace the satellites. However, this is not a convenient solution since the use of such equipment is limited to pre-defined environments. Other researchers are attempting to equip flying robots with the same kind of sensors that are commonly found on terrestrial mobile robots, i.e. range finders such as sonars or lasers [Everett, 1995; Siegwart and Nourbakhsh, 2004; Bekey, 2005; Thrun et al., 2005]. The problem with this approach is that not only do flying systems possess a very limited payload, which is very often incompatible with such sensors, but, in addition, they must survey a 3D space whereas terrestrial robots are generally satisfied with 2D scans of their surroundings. Moreover, because of their higher speed, flying robots require longer ranges of sensing, which in turn requires heavier sensors. The only known system that has been able to solve the problem of near obstacle flight using a 3D scanning laser range finder is a 100 kg helicopter equipped with a 3 kg scanning laser range finder [Scherer et al., 2007].

Even if the GPS could provide an accurate signal in near obstacle situations, the localisation information per se does not solve the collision avoidance problem. In the absence of continuously updated information concerning the surrounding obstacles, one needs to embed a very accurate 3D map of the environment in order to achieve collision-free path planning. In addition, environments are generally not completely static, and it is very difficult to incorporate into maps changes such as new buildings, cranes, etc. that could significantly disturb a UAV flying at low altitude. Apart from the problem of constructing such a map, this method would require a significant amount of memory and processing power, which may be well beyond the capability of a small flying system.

In summary, the aerial robotics community has been somehow refrained from effectively tackling the collision avoidance problem since GPS has provided an easy way around it. This problem is definitely worth getting back to in order to produce flying robots capable of flying at lower altitude or even within buildings so as to, e.g. help in search and rescue operations, provide low-altitude imagery for surveillance or mapping, measure environmental data, provide wireless communication relays, etc. Since the classical approach used in terrestrial robotics – i.e. using active distance sensors – tends to be too heavy and power consuming for flying platforms, what about turning to living systems like flies? Flies are indeed well capable of solving the problem of navigating within cluttered environments while keeping energy consumption and weight at an incredibly low level.

1.2   Flying Insects Don't Use GPS

Engineers have been able to master amazing technologies in order to fly at very high speed, relatively high in the sky. However, biological systems far outperform today's robots at tasks involving real-time perception in cluttered environments, in particular if we take energy efficiency and size into account. Based on this observation, the present book aims at identifying the biological principles that are amenable to artificial implementation in order to synthesise systems that typically require miniaturisation, energy efficiency, low-power processing and fast sensory-motor mapping.

The notion of a *biological principle* is taken in a broad meaning, ranging from individual biological features like anatomy of perceptive organs, models of information processing or behaviours, to the evolutionary process at the level of the species. The idea of applying biological principles to flying robots draws on the fields of biorobotics[2] [Chang and Gaudiano, 2000; Webb and Consi, 2001] and evolutionary robotics [Nolfi and Floreano, 2000]. These philosophical trends have in turn been inspired by the new artificial intelligence (new AI), first advocated by Brooks in the early 1980's (for a review, see Brooks, 1999) and by the seminal contribution from Braitenberg [1984]. However, when taking inspiration from biology in order to engineer artificial systems, care must be taken to avoid the pitfall of carrying out biomimicry for the sake of itself, while forgetting the primary goal, i.e. the realisation of functional autonomous robots. For instance, it would make no sense to replace efficiently engineered systems or subsystems by poorly performing bio-inspired solutions for the sole reason that they are bio-inspired. In our approach, biological inspiration will take place at different levels.

The first level concerns the selection of sensory modalities. Flies do not use GPS, but mainly low-resolution, fast and wide field-of-view (FOV) eyes, gyroscopic sensors and airspeed detectors. Interestingly, these kinds of sensors can be found in very small and low-power packages. Recent developments in MEMS[3] technology allow the measurement of strength, pressure, or inertial forces with ultra-light devices weighing only a few milligrams. Therefore, artificial sensors can easily mimic certain proprioceptive senses in flying insects. Concerning the perception of the surroundings, the only passive sensory modality that can provide useful information is vision. Active range finders such as lasers or sonars have significant drawbacks such as their inherent weight (they require an emitter and a receiver), their need to send energy into the environment, and their inability to cover a wide portion of the surroundings unless they are mounted on a mechanically scanning system. Visual sensors, on the other hand, can be extremely small, do not need to send energy into the environment, and have by essence a larger FOV. It is probable that these same considerations have driven evolution toward extensive use of vision in flying insects rather than active range finders to control their flight, avoid collisions and navigate in cluttered environments.

---

[2] Also called bio-inspired robotics or biomimetic robotics.

[3] Micro-Electro-Mechanical Systems.

The second level of bio-inspiration is related to the control system, in other words, how sensor information is processed and merged in order to provide useful motor commands. At this level, two different approaches will be explored. The first approach consists in copying flying insects in their way of processing information and behaving: controlling attitude (orientation), stabilising their course, maintaining ground clearance, and avoiding collisions. The second approach relies on artificial evolution to automatically synthesise neuromorphic controllers that map sensory signals into motor commands in order to produce a globally efficient behaviour without requiring the designer to divide it into specific sub-behaviours. In both these approaches, vision remains the core sensory modality.

However, a significant drawback with vision is the complex relationship existing between the raw signal produced by the photoreceptors and the corresponding 3D layout of the surroundings. The mainstream approach to computer vision, based on a sequence of pre-processing, segmentation, object extraction, and pattern recognition of each single image, is often incompatible with the limited processing power usually present onboard small flying robots. By taking inspiration from flying insects, this book aims at demonstrating how simple visual patterns can be directly linked to motor commands. The underlying idea is very close to the ecological approach to visual perception, first developed by Gibson [1950, 1979] and further advocated by Duchon *et al.* [1998]:

Ecological psychology (...) views animals and their environments as "inseparable pairs" that should be described at a scale relevant to the animal's behavior. So, for example, animals perceive the layout of surfaces (not the coordinates of points in space) and what the layout affords for action (not merely its three-dimensional structure). A main tenet of the ecological approach is that the optic array, the pattern of light reflected from these surfaces, provides adequate information for controlling behavior without further inferential processing or model construction. This view is called *direct perception*: The animal has direct knowledge of, and relationship to its environment as a result of natural laws.

Following this idea, no attempt will be made to, e.g. explicitly estimate distances separating the artificial eye of the flying robot and the potential obstacles. Instead, simple biological models will be used to directly link perception to action without going through complex sequences of image processing.

In summary, this book explores how principles found in insects can be applied to the design of small autonomous flying robots. This endeavor is motivated by the fact that insects have proven successful at coping with the same kinds of problems. Note that bio-inspiration could also take place at a mechanical or anatomical level. However, it is unclear whether this would improve engineered solutions. For instance, although flapping-wing mechanisms [Dickinson *et al.*, 1999; Dudley, 2000; Fry *et al.*, 2003; Lehmann, 2004] are reviewed in this book, they will not be retained as an efficient or mature-enough solution.

1.3   Proposed Approach

The research described in this book lies at the intersection of several scientific disciplines such as biology, aerodynamics, micro-engineering, micro-electronics, computer vision, and robotics. One of the main challenges therefore lies in the integration of the knowledge from various disciplines in order to develop efficient systems that will eventually be capable of autonomous flight in the presence of obstacles.

When tackling the realisation of bio-inspired flying robots, not only do the physical platforms need to be developed, but the type of behaviours they should display must be designed as must the environments in which they will be tested. Since, in the most general terms, this research field has no limits, the scope of this book has been deliberately restricted as follows.

Platforms

Recently, flying in confined indoor environments has become possible thanks to technological advances in battery technology (increase in specific energy) and miniaturisation of electrical motors [Nicoud and Zufferey, 2002]. This opportunity has opened new horizons to roboticists since small indoor flying platforms are usually less expensive, less dangerous and easier to repair in case of a crash as opposed to outdoor UAVs. However, flying indoors imposes strong constraints toward efficient system integration, minimal weight and low energy consumption. This is mainly due to the fact that in order to be at ease in an indoor environment, the inertia of the whole system needs to be kept as low as possible. With a fixed-wing aircraft, the mass is proportional to the square of the airspeed, which makes low weight essential in order to maintain the manoeuvrability in tight spaces. For instance, in order for an airplane to fly in a standard office, it needs to weigh less than 15 g or so. At such a low weight, one can easily imagine that the available payload to automate such systems is much smaller than most processing units and peripherals currently found in autonomous robots. Solving the problem of autonomous flight under such constraints therefore constitutes the core of this book.

The first step towards the creation of autonomous indoor flying robots thus consists of building platforms able to manoeuvre within confined spaces, while maintaining enough lift capability to support the required sensors and electronics. In order to progressively study and develop the required electronics and control strategies, we used a series of platforms ranging from a miniature wheeled robot to a 10-gram indoor microflyer. The purpose is to progressively increase the number of degrees of freedom, the complexity of the dynamic behaviour, and the required level of miniaturisation. The first platform is a miniature wheeled robot featuring similar electronics as subsequent flying platforms and constituting an excellent tool for fast prototyping of control strategies. The second robot is a 120 cm long indoor blimp, which naturally floats in the air and is therefore easier to control as opposed to an airplane. Due to its robustness and the fact that it does not need energy to produce lift, the blimp is well adapted to long-lasting experiments such as evolutionary runs. The last two platforms are ultra-light indoor airplanes, one weighing 30 g and the other one a mere 10 g, both flying at around 1.5 m/s.

Environments

Regarding the choice of test environments, simple geometries and textures are chosen in order to ease the characterisation of behaviours and their comparison with existing data from biologists. The test arenas are thus simple square rooms with randomly distributed black and white textures to provide contrasted visual cues. Interestingly, a striking similarity exists between our environments and the one used by some biologists to unravel the principles of insect flight control [Egelhaaf and Borst, 1993a; Srinivasan *et al.*, 1996; Tammero and Dickinson, 2002a]. The size of the arenas (from 0.6 to 15 m) is of course adapted to the natural velocity of each robot. At this early stage of bio-inspired control of indoor robots, no obstacles other than the walls themselves are considered.

Behaviours

At the behavioural level, instead of tackling an endless list of higher-level behaviours such as goal-directed navigation, homing, area coverage, food seeking, landing, etc., which themselves constitute open research topics even with robots featuring simpler dynamics, this book focuses on low-level control. An interesting way of formulating the behaviour is simply "moving forward" because, if considered over a certain period of time, this would urge the robot to remain airborne, move around, avoid collisions while implicitly requiring a series of more basic mechanisms such as attitude control, course stabilisation, and altitude control. In addition, the forward velocity is something that can easily be measured on-board the robots by means of an airspeed sensor and be used as a criteria to be optimised.

We therefore consider the ability to move forward in a collision-free manner as the first level of autonomy. Of course, specific applications or tasks would require additional behaviours on top of it, but once the first level is implemented it becomes relatively easy to add more complex behaviours on top of it using, e.g. either a subsumption or a three-layer architecture [Brooks, 1999; Bekey, 2005].

1.4   Book Organisation

Related Work (Chap. 2)

Almost no previous research has been directly aimed at insect-inspired autonomous indoor flight. However, three areas of research have been identified, which have heavily contributed to that presented in this book. The first one concerns the mechatronic design of small flying platforms, which are not yet autonomous, but may feature properties allowing for indoor flight. The second area focuses on bio-inspired vision-based navigation, which has been studied mainly on wheeled robots or in simulation. The last area is devoted to artificial evolution of vision-based control strategies.

Flying Insects (Chap. 3)

As we wish to take inspiration from flying insects, this Chapter reviews biological principles, from sensor anatomy to information processing and behaviour, that may be amenable to artificial implementation. This is not a comprehensive biological description of flying insects, but rather a pragmatic insight into selected topics from an engineering perspective.

Robotic Platforms (Chap. 4)

The platforms and tools that have been developed in order to test the proposed approach are here introduced. An overview of the four robots featuring an increasing dynamic complexity is provided along with a description of their electronics and sensors. The test arenas, adapted to the size and velocity of each robot, are also described. Additionally, the software tools allowing the interfacing and simulation of these robots is briefly presented.

Optic Flow (Chap. 5)

The detection of visual motion plays a prominent role in the behaviours of flying insects. This Chapter is therefore devoted to optic flow, its formal definition, properties, and detection. Taking into account the very limited processing power available on-board small flying robots, an efficient algorithm for estimating optic flow is proposed and characterised under real-world conditions.

Optic-flow-based Control Strategies (Chap. 6)

Taking inspiration from the models and principles described in Chapter 3 and fitting the constraints imposed by the properties of the robots presented in Chapter 4, this Chapter describes the implementation of visually-guided behaviours using optic flow. Collision avoidance and altitude control are first tested on wheels and then transferred to the indoor airplanes.

Evolved Control Strategies (Chap. 7)

One of the major problems faced by engineers that are willing to use bio-inspiration in the process of hand-crafting artificial systems is the overwhelming amount of details and varieties of biological models. An alternative approach is to rely on the principles underlying natural evolution. This so-called artificial evolution embodies the idea of transcribing Darwinian principles into artificial systems. In this Chapter, this alternative level of bio-inspiration is used to evolve neuromorphic controllers for vision-based navigation. From an engineering point of view the main advantage of relying on artificial evolution is the fact that the designer does not need to divide the desired behaviour into simple basic behaviours to be implemented into separate modules of the robot control system. After preliminary experiments on wheels, the method is applied to the blimp robot. Efficient collision avoidance and handling of critical situations are demonstrated using the same sensory modalities as in Chapter 6, namely vision, gyroscopes and airspeed sensors.

CHAPTER 2

Related Work

True creativity is characterized by a succession of acts, each dependent on the one before and suggesting the one after.

E. H. Land (1909-1991)

This Chapter reviews research efforts in the three main related domains that are micro-mechanical flying devices, bio-inspired vision-based navigation, and artificial evolution for vision-based robots. The first Section focuses on systems that are small and slow enough to be, at least potentially, capable of flying in confined environments such as houses or offices. We will see that most of them are not (yet) autonomous, either because they are too small to embed any computational power and sensors, or simply because there a light enough control system is not available. For this reason we have decided to take a more pragmatic approach to indoor flight by building upon a simpler technology that allows us to spend more efforts on control issues and miniaturisation of the embedded control-related electronics.

The two remaining Sections demonstrate how the developments presented later in this book have their roots in earlier projects, which encompass both terrestrial and aerial robots, be they real or simulated. They all share a common inspiration from biological principles as the basis of their control system. We finally present a few projects where artificial evolution has been applied to automatically create vision-based control systems.

2.1   Micromechanical Flying Devices

This Section is a review of recent efforts in the fabrication of micromechanical devices capable of flying in confined environments. We deliberately let lighter-than-air platforms (blimps) aside since their realisation is not technically challenging[1]. Outdoor micro air vehicles (MAV) as defined by DARPA[2] (see for example Mueller, 2001; Grasmeyer and Keennon, 2001; Ettinger *et al.*, 2003) are not tackled either since they are not intended for slow flight in confined areas. MAVs do indeed fly at around 15 m/s, whereas indoor aircraft are required to fly below 2 m/s in order to be able manoeuvre in offices or houses [Nicoud and Zufferey, 2002]. Nor does this Section tackle fixed-wing indoor slow flyers as two examples will be described in detail in Chapter 4.

More generally, the focus is placed on devices lighter than 15 g since we believe that heavier systems are impractical for indoor use. They tend to become noisy and dangerous for people or the surrounding objects. It is also interesting to note that developments of such lightweight flying systems have been rendered possible by the recent availability (around 2002-2003) of high discharge rate (10-20 C), high specific energy (150-200 kW/h) lithium-polymer batteries in small packages (less than 1 g).

2.1.1   Rotor-based Devices

Already in 2001, a team at Stanford University [Kroo and Kunz, 2001] developed a centimeter-scale rotorcraft using four miniature motors with 15 mm propellers. However, experiments on lift and stability were carried out on larger models and the smaller version never took off with its own battery onboard.

A few years later, Petter Muren came up with a revolutionary concept for turning helicopters into passively stable devices. This was achieved by a patented counter-rotating rotor system, which required no swash-plates

---

[1] More information concerning projects with such platforms can be found in [Zhang and Ostrowski, 1998; Planta *et al.*, 2002; van der Zwaan *et al.*, 2002; Melhuish and Welsby, 2002; da Silva Metelo and Garcia Campos, 2003; Iida, 2003; Zufferey *et al.*, 2006].

[2] The American Defense Advanced Research Projects Agency.

or collective blade control. The 3-gram Picoflyer is a good example of how this concept can be applied to produce ultralight indoor flying platforms, which can hover freely for about 1 minute (Fig. 2.1).

Figure 2.1 The remote-controlled 3-gram Picoflyer by Petter Muren. Image reprinted with permission from Petter Muren (http://www.proxflyer.com).

Almost at the same time, the Seikon Epson Corp. came up with a 12.3-gram helicopter showing off their technology in ultrasonic motors and gyroscopic sensors (Fig. 2.2). Two ultra-thin, ultrasonic motors driving two contra-rotating propellers allow for a flight time of 3 minutes. An image sensor unit could capture and transmit images via a Bluetooth wireless connection to an off-board monitor.

2.1.2   Flapping-wing Devices

Another research direction deserving increasing attention concerns flapping-wing devices. A team at Caltech in collaboration with Aeroviron-ment[TM] developed the first remote-controlled, battery-powered, flapping-wing micro aircraft [Pornsin-Sirirak *et al.*, 2001]. This 12-gram device with a 20 cm wingspan has an autonomy of approximately 6 minutes when pow-ered with a lithium-polymer battery. However, the Microbat tended to fly fast and was therefore only demonstrated in outdoor environments.

Figure 2.2 The 12.3-gram uFR-II helicopter from Epson. Image reproduced with permission from Seiko Epson Corporation (http://www.epson.co.jp).

Figure 2.3 The Naval Postgraduate School 14-gram biplane flapping thruster. Reprinted with permission from Dr Kevin Jones.

More recently, Jones *et al.* [2004] engineered a small radio-controlled device propelled by a novel biplane configuration of flapping wings moving up and down in counter-phase (Fig. 2.3). The symmetry of the flapping wings emulates a single wing flapping in ground effect, producing a better performance, while providing an aerodynamically and mechanically balanced system. The downstream placement of the flapping wings helps prevent flow separation over the main wing, allowing the aircraft to fly efficiently at very low speeds with high angles of attack without stall. The 14-gram model has demonstrated stable flight at speeds between 2 and 5 m/s.

Probably the most successful flapping microflyer to date is the DelFly [Lentink, 2007], which has been developed in the Netherlands by TU Delft, Wageningen University and Ruijsink Dynamic Engineering. It has four flexible, sail-like, wings placed in a bi-plane configuration and is powered by a single electric motor (Fig. 2.4). The aircraft can hover almost motionlessly in one spot as well as fly at considerable speed. The latest version weighs 15 to 21 g (depending on the presence or not of an embedded

Figure 2.4 The 15-gram flapping-wing DelFly is capable of both hovering and fast forward flight. Reprinted with permission from Dr David Lentink (http://www.delfly.nl).

Figure 2.5 The artist's conception (credits Q. Gan, UC Berkeley) and a preliminary version of the micromechanical flying insect (MFI). Reprinted with permission from Prof. Ron Fearing, UC Berkeley.

camera) and can fly for more than 15 minutes. Although it is not able to fly autonomously while avoiding collisions, DelFly can be equipped with a small camera that sends images to an offboard computer to, e.g. detect targets. Motivated by the amazing flight capabilities of Delfly, many other flapping wings are being developed, of which some were presented at the International Symposium on Flying Insects and Robots in Switzerland [Floreano *et al.*, 2007].

On an even smaller scale, Ron Fearing's team is attempting to create a micro flying robot (Fig. 2.5) that replicates the wing mechanics and dynamics of a fly [Fearing *et al.*, 2000]. The planned weight of the final device is approximately 100 mg for a 25 mm wingspan. Piezoelectric actuators are used for flapping and rotating the wings at about 150 Hz. Energy is planned to be supplied by lithium-polymer batteries charged by three miniature solar panels. So far, a single wing on a test rig has generated an average lift of approximately 0.5 mN while linked to an off-board power supply [Avadhanula *et al.*, 2003]. Two of these wings would be sufficient to lift a 100 mg device. The same team is also working on a bio-mimetic sensor suite for attitude control [Wu *et al.*, 2003], but no test in flight has been reported so far.

Although these flying devices constitute remarkable micro-mechatronic developments, none of them includes a control system allowing for autonomous operation in confined environments.

2.2   Bio-inspired Vision-based Robots

In the early 90's, research on biomimetic vision-based navigation was mainly carried out on wheeled robots. Although some researchers have shown interest in higher level behaviours such as searching, aiming and navigating by using topological landmarks, etc. (for a review see Franz and Mallot, 2000), we focus here on the lower level, which is mainly collision avoidance. More recently, similar approaches have been applied to aerial robotics and we will see that only subproblems have been solved in this area. A common aspect of all these robots is that they use optic flow (see Chap. 5) as their main sensory input for controlling their movements.

2.2.1   Wheeled Robots

Franceschini and his team at CNRS in Marseille, France, have spent several years studying the morphological and neurological aspects of the visual system of flies and their way of detecting optic flow (for a review, see Franceschini, 2004). In order to test their hypotheses on how flies use optic flow, the team built an analog electronic circuit modeled upon the neural circuitry of the fly brain and interfaced it with a circular array of photoreceptors on a 12-kg wheeled robot (Fig. 2.6). The so-called "robot mouche" was capable of approaching a goal while avoiding obstacles in its path [Pichon *et al.*, 1990; Franceschini *et al.*, 1992]. The obstacles were characterised by higher contrasts with respect to a uniform background. The robot used a series of straight motions and fast rotations to achieve a collision-free navigation.

Although some preliminary results in vision-based collision avoidance have been obtained with a gantry robot by Nelson and Aloimonos [1989] most of the work on biomimetic vision-based robots has followed the realisation of the "robot mouche". Another key player in this domain is Srinivasan and his team at the Australian National University in Canberra. They have performed an extensive set of experiments to understand the visual performance of honeybees and have tested the resulting models on robots (for reviews, see Srinivasan *et al.*, 1997, 1998). For example, they demonstrated that honeybees regulate their direction of flight by balancing the optic flow on their two eyes [Srinivasan *et al.*, 1996]. This mechanism was then demonstrated on a wheeled robot equipped with a camera and two mirrors (Fig. 2.7upper) capturing images of the lateral walls and transmitting them to a desktop computer where an algorithm attempted to balance the optic flow in the two lateral views by steering the robot accordingly [Weber *et al.*, 1997]. In the same team, Sobey [1994] implemented an algorithm inspired by insect flight to drive a vision-based robot (Fig. 2.7lower) in cluttered environments. The algorithm related the position of the camera, the speed of the robot, and the measured optic flow during translational motions in order to estimate distances from objects and steer accordingly.

Several other groups have explored the use of insect visual control systems as models for wheeled robot navigation, would it be for collision avoidance in cluttered environments [Duchon and Warren, 1994; Lewis, 1998] or corridor following [Coombs *et al.*, 1995; Santos-Victor *et al.*, 1995]. In some of these robots, active camera mechanisms have been employed for stabilising their gaze in order to cancel spurious optic-flow introduced by self-rotation (a processed called *derotation*, see Chap. 5).

Figure 2.6 The "robot mouche" has a visual system composed of a compound eye (visible at half-height) for obstacle avoidance, and a target seeker (visible on top) for detecting the light source serving as a goal. Reprinted with permission from Dr Nicolas Franceschini.

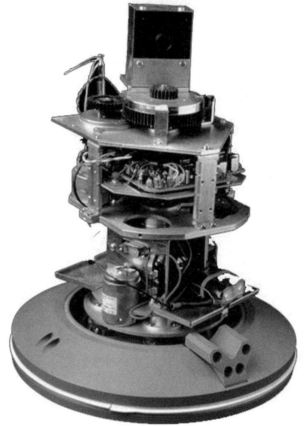

Figure 2.7   (Upper) The corridor-following robot by Srinivasan's team. Reprinted with permission from Prof. Mandyam V. Srinivasan. (Lower) The obstacle-avoiding robot by Srinivasan's team. Reprinted with permission from Prof. Mandyam V. Srinivasan.

However, all of these robots rely on the fact that they are in contact with a flat surface in order to infer or control their self-motion through wheel encoders. Since flying robots have no contact with the ground, the proposed approaches cannot be directly applied to flying devices. Furthermore, the tight weight budget precludes active camera mechanisms for gaze stabilisation. It is also worth mentioning that all the above wheeled robots, with the sole exception of the "robot mouche", used off-board image processing and were therefore not self-contained autonomous systems.

2.2.2   Aerial Robots

A few experiments on optic-flow-based navigation have been carried out on blimps. Iida and colleagues have demonstrated visual odometry and course stabilisation [Iida and Lambrinos, 2000; Iida, 2001, 2003] using such a platform equipped with an omnidirectional camera (Fig. 2.8) down-streaming images to an off-board computer for optic-flow estimation. Planta et al. [2002] have presented a blimp using an off-board neural controller for course and altitude stabilisation in a rectangular arena equipped with regular checkerboard patterns. However, altitude control produced very poor results. Although these projects were not directly aimed at collision avoidance, they are worth mentioning since they are among the first realisations of optic-flow-based indoor flying robots.

Specific studies on altitude control have been conducted by Franceschini's group, first in simulation [Mura and Franceschini, 1994], and more recently with tethered helicopters (Fig. 2.9; Netter and Franceschini, 2002; Ruffier and Franceschini, 2004). Although the control was performed off-board, the viability of regulating the altitude of a small helicopter using the amount of ventral optic flow as detected by a minimalist vision system (only 2 photoreceptors) could be demonstrated. The regulation system did not even need to know the velocity of the aircraft. Since these helicopters were tethered, the number of degrees of freedom were deliberately limited to 3 and the pitch angle could directly be controlled by means of a servomotor mounted at the articulation between the boom and the aircraft. The knowledge of the absolute pitch angle made it possible to ensure the vertical orientation of the optic-flow detector when the rotorcraft was tilted fore

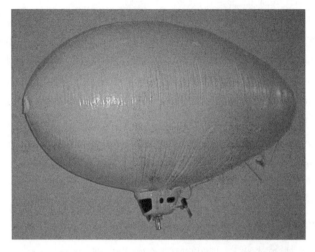

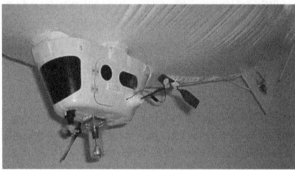

Figure 2.8 (Upper) Melissa is an indoor blimp for visual odometry experiments. (Lower) Closeup showing the gondola and the omnidirectional vision system. Reprinted with permission from Dr Fumiya Iida.

and aft to modulate its velocity. On a free-flying system, it would not be trivial to ensure the vertical orientation of a sensor at all time.

In an attempt at using optic-flow to control the altitude of a free-flying UAV, Chahl *et al.* [2004] took inspiration from the landing strategy of honeybees [Srinivasan *et al.*, 2000] to regulate the pitch angle using ventral optic-flow during descent. However, real world experiments produced very limited results, mainly because of the spurious optic-flow introduced by corrective pitching movements (no derotation). In a later experiment, Thakoor *et al.* [2004] achieved altitude control over a flat desert ground (Fig. 2.10) using a mouse sensor as an optic-flow detector. However, no detailed data has been provided regarding the functionality and robustness of the system.

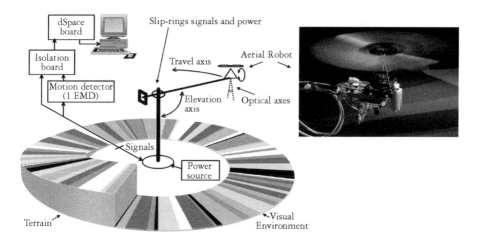

Figure 2.9 The tethered helicopter used for the optic-flow-based altitude control study. Reprinted with permission from Dr Nicolas Franceschini and Dr Franck Ruffier. Picture copyright H. Raguet and Photothèque CNRS, Paris.

In order to test a model of collision avoidance in flies [Tammero and Dickinson, 2002a], Reiser and Dickinson [2003] set up an experiment with a robotic gantry (Fig. 2.11) emulating a fly's motion in a randomly textured circular arena. This experiment successfully demonstrated a robust collision avoidance. However, the experiment only considered motion in a 2D plane.

Figure 2.10 The UAV equipped with a ventral optical mouse sensor for altitude control. Reprinted from Thakoor et al. [2004], copyright IEEE.

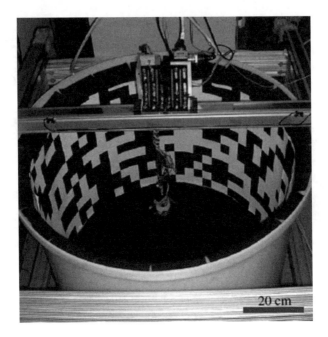

Figure 2.11 The gantry system is capable of moving a wide-FOV camera through the arena. Reprinted from Reiser and Dickinson [2003, figure 3] with permission from The Royal Society.

Another significant body of work entirely conducted in simulation [Neumann and Bülthoff, 2001, 2002] demonstrated full 3D, vision-based navigation (Fig. 2.12). The attitude of the agent was maintained level using

Figure 2.12 Closed-loop autonomous flight control using fly-inspired optic flow to avoid obstacle and light gradient to keep attitude level at all time [Neumann and Bülthoff, 2002]. Reprinted with permission from Prof. Heinrich H. Bülthoff.

the light intensity gradient; course stabilisation, obstacle avoidance and altitude control were based on optic flow. However, the dynamics of the simulated agent was minimalist (not representative of a real flying robot) and the environment featured a well-defined light intensity gradient, which is not always available in real-world conditions, especially when flying close to obstacles or indoors.

More recently, Muratet *et al.* [2005] developed an efficient optic-flow-based control strategy for collision avoidance with a simulated helicopter flying in urban canyons. However, this work in simulation relied on a full-featured autopilot (with GPS, inertial measurement unit, and altitude sensor) as its low-level flight controller and made use of a relatively high resolution camera. These components are likely to be too heavy when it comes to the reality of ultra-light flying robots.

The attempts at automating real free-flying UAVs using bio-inspired vision are quite limited. Barrows *et al.* [2001] have reported on preliminary experiments on lateral obstacle avoidance in a gymnasium with a model glider carrying a 25-gram optic-flow sensor. Although no data supporting the described results are provided, a video shows the glider steering away from a wall when tossed toward it at a shallow angle. A further experiment with a 1-meter wingspanned aircraft Barrows *et al.* [2002] was performed outdoors. The purpose was essentially to demonstrate altitude control with a ventral optic-flow sensor. A simple (on/off) altitude control law managed to maintain the aircraft airborne for 15 minutes, during which 3 failures occurred where the human pilot had to rescue the aircraft due to it dropping too close to the ground. More recently, Green *et al.* [2004] carried out an experiment on lateral obstacle avoidance with an indoor aircraft equipped with a laterally-mounted 4.8-gram optic-flow sensor (Fig. 2.13). A single trial, in which the aircraft avoided a basketball net is described and illustrated with video screen-shots. Since merely one sensor was used, the aircraft could detect obstacles only on one side. Although these early experiments by Barrows, Green and colleagues are remarkable, no continuous collision-free flight in confined environments has been reported so far. Furthermore, no specific attention has been made to derotate the optic-flow signals. The authors assumed – more or less implicitly – that rotational components of optic flow arising from changes in aircraft orientation are smaller than the translational component. However, this assumption does not usually hold true (in particular when the robot is required to actively avoid obstacles) and this issue deserves more careful attention. Finally, no frontal collision avoidance experiments have thus far been described.

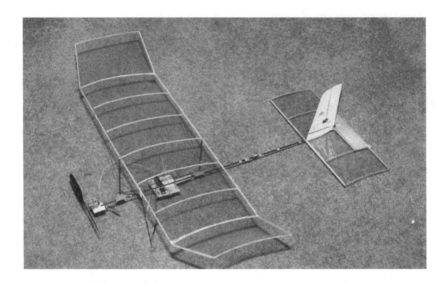

Figure 2.13 Indoor flyer (about 30 g) with a single lateral optic-flow detector (4.8 g). Reprinted with permission from Prof. Paul Oh and Dr Bill Green.

More recently, Griffiths *et al.* [2007] have used optic-flow mouse sensors as complementary distance sensors navigational aids for an aerial platform (Fig. 2.14) in mountainous canyons. The robot is fully equipped with an inertial measurement unit (IMU) and GPS. It computes the optimal 3D path based on an *a priori* 3D map of the environment. In order to be able to react to unforeseen obstacles on the computed nominal path, it uses the frontal laser range finder and two lateral optical mouse sensors. This robot has demonstrated low altitude flight in a natural canyon while the mouse sensors provided a tendency towards the center when the nominal path was deliberately biased towards one or the other side of the canyon. Although no data showing the accuracy of measurements are provided, the experiment demonstrated that by carefully derotating optic-flow measurements from the mouse sensors, such information can be used to estimate rather large distances in outdoor environments.

Figure 2.14 The 1.5-m-wingspanned platform used for autonomous flight in canyons. The square hole in the center is the Opti-Logic RS400 laser range-finder (400 m range, 170 g, 1.8 W), and the round holes are for Agilent ADNS-2610 optical mouse sensors. Courtesy of the BYU Magicc Lab.

Hrabar *et al.* [2005], also used lateral optic-flow to enable a large helicopter (Fig. 2.15) to center among obstacles outdoors, while another kind of distance sensor (stereo vision) was utilized to avoid frontal obstacles. However, in these last two projects the vision sensors were by no means used as primary sensors and the control system relied mainly on a classical and relatively bulky autopilot.

In all the reviewed projects, the vision-based control system only helps with or solves part of the problem of close-obstacle, collision-free navigation. In addition, none of the proposed embedded electrics would fit a 10-gram robot.

2.3   Evolution of Vision-based Navigation

Instead of hand-crafting robot controllers based on biological principles, an alternative approach consists in using genetic algorithms[3] (GAs). When

---

[3] Search procedure based on the mechanisms of natural selection [Goldberg, 1989].

Figure 2.15 The USC Autonomous Helicopter platform (AVATAR) equipped with two wide-FOV lateral cameras. Reprinted with permission from Dr Stefan Hrabar.

applied to the design of robot controllers, this method is called evolutionary robotics (ER) and goes as follows [Nolfi and Floreano, 2000]:

> An initial population of different artificial chromosomes, each encoding the control system (and sometimes the morphology) of a robot, are randomly created and put in the environment. Each robot (physical or simulated) is then let free to act (move, look around, manipulate) according to a genetically specified controller while its performance on various tasks is automatically evaluated. The fittest robots are allowed to reproduce by generating copies of their genotypes with the addition of changes introduced by some genetic operators (e.g. mutations, crossover, duplication). This process is repeated for a number of generations until an individual is born which satisfies the performance criterion (fitness function) set by the experimenter.

Certain ER experiments have already demonstrated successful results at evolving vision-based robots to navigate. Those related to collision avoidance are briefly reviewed in this Section.

At the Max-Plank Institute in Tübingen, Huber *et al.* [1996] have carried out a set of experiments where a simulated agent evolved its visual sensor orientations and sensory-motor coupling. The task of the agent was to navigate as far as possible in a corridor-like environment with a few perpendicular obstacles. Four photodetectors were brought together to compose two elementary motion detectors (see Chap. 3), one on each side of the agent. The simple sensory-motor architecture was inspired from Braitenberg [1984]. Despite their minimalist sensory system, the autonomous agents successfully adapted to the task during artificial evolution. The best evolved individuals had a sensor orientation and a sensory-motor coupling suitable for collision avoidance.

Going one step further, Neumann *et al.* [1997] showed that the same approach could be applied to simulated aerial agents. The minimalist flying system was equipped with two horizontal and two vertical elementary motion detectors and evolved in the same kind of textured corridor. Although the agents developed effective behaviours to avoid horizontal and vertical obstacles, such results are only of limited interest when it comes to physical flying robots since the simulated agents featured very basic dynamics and had no freedom around their pitch and roll axes. Moreover, the visual input was probably too perfect and noise-free to be representative of real-world conditions[4].

At the Swiss Federal Institute of Technology in Lausanne (EPFL), Floreano and Mattiussi [2001] have carried out experiments where a small wheeled robot evolved the ability to navigate in a randomly textured environment. The robot was equipped with a 1D camera composed of 16 pixels with a $36°$ FOV as its only sensor. Evolution could relatively quickly find functional neuromorphic controllers capable of navigating in the environment without hitting the walls, and this by using a very simple genetic encoding and fitness function. Note that unlike the experiments by Huber and Neumann, this approach did not explicitly use optic flow, but rather

---

[4] Other authors have evolved terrestrial vision-based robots in simulation (for example, Cliff and Miller, 1996; Cliff *et al.*, 1997), but the chosen tasks (pursuit and evasion) are not directly related to the ones tackled in this book. The same team has also worked with a gantry robot for real-world visually-guided behaviours such as shape discrimination [Harvey *et al.*, 1994].

raw vision. The visual input was simply preprocessed with a spatial high-pass filter before feeding a general purpose neural network and the sensory morphology was not concurrently evolved with the controller architecture.

Another set of experiments [Marocco and Floreano, 2002; Floreano *et al.*, 2004], both in simulation and with a real robot, explored the evolution of active visual mechanisms allowing evolved controllers to decide where to look while they were navigating in their environment. Although those experiments yielded interesting results, this approach was discarded for our application since an active camera mechanism is too heavy for the desired aerial robots.

2.4   Conclusion

Many groups have been or are still working on the development of micromecanical devices capable of flying in confined environments. However, this field is still in its infancy and will require advances in small-scale and low Reynolds aerodynamics as well as micro actuators and small-scale, high specific-energy batteries. In this book, a pragmatic approach is taken using a series of platforms ranging from wheeled, to buoyant, to fixed-wing vehicles. Although it was developed 3 years earlier, our 10-gram microflyer (Chap. 4) can compete in manoeuvrability and endurance with the most recent flapping-wing and rotor-based platforms. Nevertheless, a fixed-wing platform is easier to build and can better withstand the crashes that will occur during the development process.

On the control side, the bio-inspired vision-based robots developed up until now have been incapable of demonstrating full 3D autonomy in confined environments. We will show how this is possible while keeping the embedded control system at a weight below 5 g using mostly off-the-shelf components. The result naturally paves the way towards automating the other micromechanical flying devices presented above as well as their successors.

CHAPTER 3

Flying Insects

The best model of a cat for biologists is another or better, the same cat.

N. Wiener (1894-1964)

This Chapter reviews biological principles related to flight control in insects. In the search for biological principles that are portable to artificial implementation in lightweight flying robots, the review is organised into three levels of analysis that are relevant for the control of both robots and animals: sensors (perceptive organs), information processing, and behaviour.

This book has its main interest in flying insects since they face constraints that are very similar to those encountered by small aerial robots, notably a minimal power consumption, ultra-low weight, and the control of fast motion in real time. Relying on animal taxonomy, we first briefly discuss which insects are the most interesting in our endeavour and why.

3.1   Which Flying Insects?

The animal kingdom is divided into phyla, among which the arthropods are composed of four classes, one of those being the insects. Arthropods are invertebrate animals possessing an exoskeleton, a segmented body, and jointed legs. The compound eyes of arthropods are built quite differently as opposed to eyes of vertebrates. They are made up of repeated units called *ommatidia*, each of which functions as a separate visual receptor with its own lens (see Sect. 3.2.1).

Among arthropods, the most successful flying animals are found in the insect class, which is itself divided into orders such as Diptera (flies and mosquitoes), Hymenoptera (bees), Orthoptera (grasshoppers), Coleoptera (beetles), Lepidoptera (butterflies), Isoptera (termites), Hemiptera (true bugs), etc. This book focuses mainly on Diptera and Hymenoptera, not only because flies and bees are generally considered to be the best flyers, but also because a few species of these two orders, namely the blowflies (Calliphora), the houseflies (Musca), the fruitflies (Drosophila), and the honeybees (Apis), have been extensively studied by biologists (Fig. 3.1). Almost all insects have two pairs of wings, whereas Diptera feature only one pair. Their hind wings have been transformed through evolution into tiny club-shaped mechanosensors, named *halteres*, which provide gyroscopic information (see Sect. 3.2.2).

Figure 3.1 An example of highly capable and thoroughly studied flying insect: the blowfly Calliphora. Copyright by Flagstaffotos.

The sensory and nervous systems of flies have been analysed for decades, which has resulted in a wealth of electrophysiological data, models of information processing and behavioural descriptions. For example, many neurons in the fly's brain have been linked to specific visually-guided behaviours

[Egelhaaf and Borst, 1993a]. Although honeybees are capable of solving a great variety of visually controlled tasks [Srinivasan *et al.*, 1996, 2000], comparatively little is known about the underlying neuronal basis. However, interesting models of visually guided strategies are available from behavioural studies.

Perception and action are part of a single closed loop rather than separate entities, but subdividing this loop into three levels helps to highlight the possibilities of artificial implementation. At the first level, the anatomical description of flying insects can be a source of inspiration for constructing a robot. Although this book is not oriented toward mechanical biomimetism, the choice of sensor modalities available on our robots (Chap. 4) is based on perceptive organs used by insects. At the second level, models of biological information processing will guide us in the design of sensory signal processing (Chap. 5). Mainly related to vision, these models have been essentially produced from neurophysiological studies or from behavioural experiments with tethered animals (see, e.g. Egelhaaf and Borst, 1993a). At the third level, the study of free-flight behaviour (ethology) provides significant insight into how insects steer in their environments and manage to take full advantage of their sensor characteristics by using specific, stereotyped movements. Similar behaviours are implemented in our robots (Chap. 6).

In the remainder of this Chapter, existing descriptions of biological principles are reviewed following the same three levels. However, this brief overview is not an extensive detailing of the biology of flying insects. Only models relevant to the basic behaviours described in the introduction (e.g. attitude control, course stabilisation, collision avoidance and altitude control) and that are potentially useful for small flying robots are presented.

3.2   Sensor Suite for Flight Control

Insects have sense organs that allow them to see, smell, taste, hear and touch their environment [Chapman, 1998]. In this Section, we focus on the sensors that are known to play an important role in flight control. Whereas flying insects use many sensor modalities, their behaviour is mainly dominated by visual control. They use visual feedback to stabilise their flight [Egelhaaf and Borst, 1993b], control their flight speed [Srinivasan *et al.*, 1996; Srinivasan and Zhang, 2000; Baird *et al.*, 2006], perceive depth [Srinivasan *et al.*, 1991; Tammero and Dickinson, 2002a], track objects [Egelhaaf and Borst, 1993b], land [Borst, 1990; Srinivasan *et al.*, 2000], measure self-motion [Krapp and Hengstenberg, 1996; Krapp, 2000] and estimate travelled distances [Srinivasan *et al.*, 2000]. The compound eye is therefore presented first together with the ocelli, a set of three photosensitive organs arranged in a triangle on the dorsal part of the head (Fig. 3.2). Subsequently, the gyroscope of Diptera, the halteres, is described since it is believed to provide the vestibular sense to flies. The last Section of this review is devoted to other mechanosensors such as the antennas and hairs, that are likely to play an important role in flight control, for example for sensing the airflow around the body.

3.2.1   Vision

Flying insects (and arthropods in general) have two large compound eyes [Chapman, 1998, p. 587] that occupy most of their head (Fig. 3.2). Each

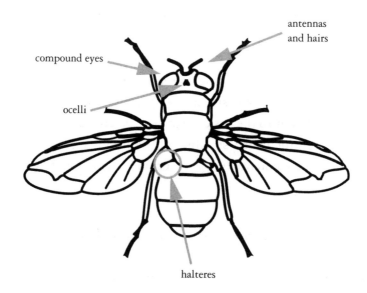

Figure 3.2 The most important perceptive organs related to flight control: the large compound eyes (and the ocelli), the halteres, and the antennas and hairs.

eye is made up of tiny hexagonal lenses, also called facets, that fit together
like the cells of a honeycomb (Fig. 3.3). Each lens admits a small part
of the total scene that the insect visualises. All the parts combine together and form the whole picture. Underlying the lens is a small tube,

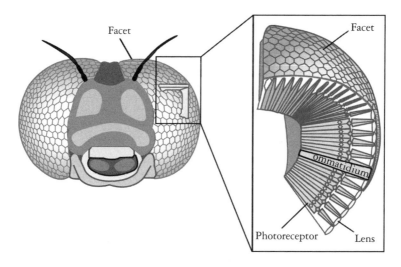

Figure 3.3 The compound eyes of flying insects. The compound eyes are made
up of repeating units, the ommatidia, each of which functions as a separate visual
receptor. Each ommatidium consists of a lens (the front surface of which makes up
a single facet), a transparent crystalline cone, light-sensitive visual cells arranged
in a radial pattern, and pigment cells that separate the ommatidium from its
neighbours.

the ommatidium, containing several photosensitive cells (for details, see
Franceschini, 1975). For the sake of simplicity, we assume in this book
that one ommatidium corresponds to one viewing direction and thus to
one pixel, although different kinds of compound eyes exist with different
arrangements [Land, 1997]. In insects, the number of ommatidia varies
from about 6 in some worker ants up to $30'000$ in dragonflies. In Diptera,
this range is smaller and varies from 700 in the fruitfly to 6000 ommatidia
per eye in the blowfly, covering roughly 85% of the visual field (maximum
possible solid angle whose apex is located at the center of the eye). Taking
the square root of the number of ommatidia, the eye of the fruitfly is thus
roughly equivalent to a 26×26 pixel array covering one visual hemisphere,
which is much less than in state-of-the-art artificial vision sensors (Fig. 3.4).

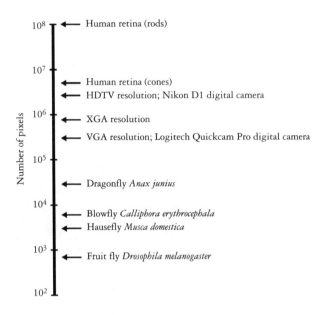

Figure 3.4 The number of pixels in artificial and biological vision systems (single eyes). The number of pixels in the eyes of flying insects is orders of magnitude lower than in current silicon imagers [Harrison, 2000; Land, 1997].

To compare the resolution power of vision systems, one has to consider not only the number of pixels but also the covered field, or more precisely the ratio of the field of view (FOV) to the number of pixels. According to Land [1997], many flying insects have an interommatidial angle in the range of 1-5° (blowfly: 1.1°, housefly: 2.5°, fruitfly: 5°), and this angle corresponds to the visual space that a single ommatidia is able to sample (acceptance angle). The best resolving power achievable by the fly's eye is thus 60 times inferior of that of a human eye. However, the compound eye configuration permits a much wider FOV because of the juxtaposition of small tubes aimed at divergent orientations instead of a single lens and a focal plane.[1] Indeed, flies can see in almost every direction except in the blind spot caused by their own body.

It is remarkable that flies are capable of such impressive flight control when considering the low-resolution of their eyes, which is a consequence

---

[1] See [Neumann, 2002] for a nice reconstruction of what flies see.

of their compound design. Moreover, because of their eye arrangement they cannot estimate distances from stereo-vision or focus, as outlined by [Srinivasan *et al.*, 1999]:

> Unlike vertebrates, insects have immobile eyes with fixed-focus optics. Thus, they cannot infer the distance of an object from the extent to which the directions of gaze must converge to view the object, or by monitoring the refractive power that is required to bring the image of the object into focus on the retina. Furthermore, compared with human eyes, the eyes of insects are positioned much closer together, and possess inferior spatial acuity. Therefore the precision with which insects could estimate the range of an object through binocular stereopsis would be much poorer and restricted to relatively small distances, even if they possessed the requisite neural apparatus.

However, fly vision greatly exceeds human vision in the temporal domain. Human vision is sensitive to temporal frequencies up to 20 Hz, whereas ommatidia respond to temporal frequencies as high as 200-300 Hz [Dudley, 2000, p. 206]. This allows flying insects to be very good at detecting changes in the visual field and especially *optic flow* (see Sect. 3.3).

In addition to their compound eyes, numerous insects have three simple photoreceptors, called *ocelli*. These ocelli are set in the form of a triangle between the compound eyes (Fig. 3.2). Since they are unfocused, they cannot form images. Rather, they are used to measure brightness and are thought to contribute to the dorsal light response where the fly aligns its head with sources of brightness [Schuppe and Hengstenberg, 1993]. Therefore, ocelli might be used to provide information about the location of the horizon in outdoor environments.

3.2.2   Vestibular Sense

In many fast-moving animals inputs from mechanosensory organs (such as the labyrinth in the ears of vertebrates) contribute to compensatory reactions, and are generally faster than what can be detected through the visual system independently of lighting conditions [Nalbach and Hengstenberg, 1994]. Diptera possess a remarkable organ for measuring angular velocities [Chapman, 1998, p.196]. Rotations of their body are per-

ceived through the halteres (Fig. 3.2, also visible in Figure 3.1a), which
have evolved by the transformation of the hind wings into tiny club-shaped
organs that oscillate during flight in antiphase with the wings [Nalbach,
1993]. These mechanosensors measure angular velocity by sensing the periodic Coriolis forces that act upon the oscillating haltere when the fly rotates
[Hengstenberg, 1991]. Coriolis effects are inertial forces acting on bodies
moving in a non-inertial (rotating) reference frame. The forces measured by
the halteres are proportional to the angular velocity of the fly's body.

According to Dickinson [1999] haltere feedback has two roles. The first
one is gaze stabilisation:

> One important role of the haltere is to stabilize the position of
> the head during flight by providing feedback to the neck motor
> system. (…) Nalbach and Hengstenberg demonstrated that the
> blowfly, Calliphora erythrocephala, discriminates among oscilla-
> tions about the yaw, pitch and roll axes and uses this information to
> make appropriate compensatory adjustments in head position (…);
> [Nalbach, 1993; Nalbach and Hengstenberg, 1994]. Such reflexes
> probably act to minimize retinal slip during flight, thereby stabil-
> ising the image of the external world and increasing the accuracy
> with which the visual system encodes motion.

The second role of the halteres consists in direct flight stabilisation:

> Although the role of the haltere in stabilising gaze may be impor-
> tant, a more essential and immediate role of the haltere is to pro-
> vide rapid feedback to wing-steering muscles to stabilize aerody-
> namic force moments.

More recently, gyroscopic sense has also been discovered in insects that do
not possess halteres. Sane et al. [2007] have shown that the antennas of
wasps could also vibrate and sense Coriolis forces much like the halteres in
Dipteras.

In summary, flight stabilisation in flies – and probably other flying
insects – is ensured by a combination of visual and vestibular senses and
both sensory modalities are of interest for the realisation of artificial systems.

3.2.3 Airflow Sensing and Other Mechanosensors

Although less thoroughly studied, it is likely that flying insects integrate information from other perceptive organs to control their flight. One of those are the bell-shaped campaniform sensilla [Chapman, 1998, p. 195] that act as strain gauges. About 335 sensilla are indeed located at the haltere base in order to detect Coriolis forces [Harrison, 2000]. Campaniform sensilla are also present on the wings allowing a perception of wing load [Hengstenberg, 1991].

Aerodynamically induced bending in external structures such as antennas potentially provides information concerning the changing speed and direction of flight [Dudley, 2000]. As noted by [Hausen and Egelhaaf, 1989], antennas are likely to participate in the mechanosensory feedback. Flying insects are also equipped with a multitude of tiny bristles (Fig. 3.5) that could help in controlling flight by providing information about air movements and changes in air pressure. In an experiment on the interaction between vision and haltere feedback, Sherman and Dickinson [2004] noted:

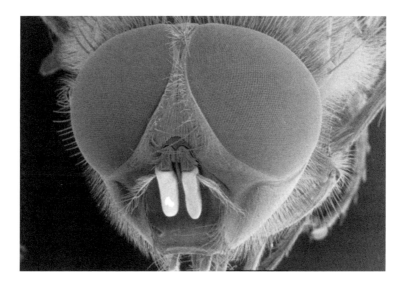

Figure 3.5 The head of a common blowfly. This image shows the hairy nature of these insects. Copyright of The University of Bath UK and reprinted with permission.

Posternal hairs on the neck, and wing campaniform sensilla could
contribute to both the basic response to mechanical oscillation and
the attenuation of the visual reflex during concurrent presentation.

As described in his thesis, Harrison [2000] also presumes that flies are able
to estimate linear accelerations through proprioceptive sensors that equip
the legs and neck, and that are able to measure position and strain.

What should be retained from this brief description of other mechanosensors that are found all around their body is that insects are very likely
to have a good perception of airflow and thus airspeed. Therefore, it may
be interesting to equip flying robots with airflow sensors, which should not
necessarily be linear.

3.3   Information Processing

Among the sensory modalities that are involved in insect flight control,
visual cues exert a predominant influence on orientation and stability. This
present Section thus focuses on vision processing. The importance of
vision for flight is underlined by the relative size of the brain region
dedicated to the processing of afferent optical information. The visual system of flies has been investigated extensively by means of behavioural experiments and by applying neuroanatomical and electrophysiological techniques. Both the behaviour and its underlying neuronal basis can sometimes be studied quantitatively in the same biological system under similar
stimulus conditions [Krapp, 2000]. Moreover, the neuronal system of flying
insects is far simpler than that of vertebrates, ensuring biologists a better
chance to link behaviour to single neuron activity. The fact that the direct
neuronal chain between the eye and the flight muscles consists of only 6-7
cells [Hausen and Egelhaaf, 1989] further illustrates the simplicity of the
underlying processing. When electrophysiological investigations are not
possible – e.g. because of the small size of some neurons – it is sometimes
still possible to deduce mathematical models of the functioning of neuronal
circuits by recording from downstream neurons.

3.3.1 Optic Lobes

The optic lobes (i.e. peripheral parts of the nervous system in the head, see Figure 3.6) of flies are organised into three aggregates of neurones (also called ganglia or neuropils): the *lamina*, the *medulla*, and the *lobula complex* (lobula and lobula plate), corresponding to three centers of vision processing. The retinotopic[(2)] organisation is maintained through the first two neuropils down to the third one, the lobula, where a massive spatial integration occurs and information from very different viewing directions are pooled together:

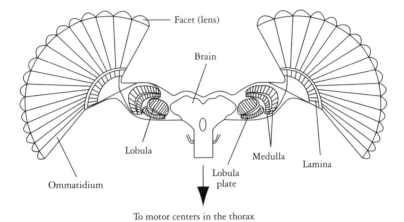

Figure 3.6 A schematic representation of the fly's visual and central nervous system (cross section through the fly's brain). Photoreceptor signals are transmitted to the lamina, which accentuates temporal changes. A retinotopic arrangement is maintained through the medulla. The lobula plate is made up of wide-field, motion-sensitive tangential neurons that send information to the controlateral optic lobe as well as to the thoracic ganglia, which control the wings. Adapted from Strausfeld [1989].

- The *lamina* lies just beneath the receptor layer of the eye and receives direct input from photoreceptors. The neurons in this ganglion act as high-pass filters by amplifying temporal changes. They also provide a gain control functionality thus ensuring a quick adaptation to varia-

---

[(2)] The neighbourhood is respected, i.e. neurons connected to neighbouring ommatidia are next to each other.

tions in background light intensity. Axons from the lamina invert the
image from front to back while projecting to the medulla.

- Cells in the *medulla* are extremely small and difficult to record from
(see, e.g. Douglass and Strausfeld, 1996). However, behavioural experiments suggest that local optic-flow detection occurs at this level (see
Sect. 3.3.2). The retinotopic organisation is still present in this second ganglion and there are about 50 neurons per ommatidium. The
medulla then sends information to the lobula complex.

- The third optic ganglion, the *lobula complex*, is the locus of massive spatial convergence. Information from several thousand photoreceptors,
preprocessed by the two previous ganglia, converges onto a mere 60
cells in the lobula plate [Hausen and Egelhaaf, 1989]. These so-called
tangential cells (or LPTC for Lobular Plate Tangential Cells) have broad
dendritic trees that receive synaptic inputs from large regions of the
medulla, resulting in large visual receptive fields (see Sect. 3.3.3). The
lobula complex projects to higher brain centers and to descending neurons that carry information to motor centers in the thoracic ganglia.

From an engineering perspective, the lamina provides basic functionalities of image preprocessing such as temporal and spatial high-pass filtering
as well as an adaptation to background light. Although generally useful,
such functionalities will not be further described nor implemented in our
artificial systems because of the relative visual simplicity of our test environments (Sect. 4.4). The two following ganglia, however, are more interesting since they feature typical properties used by flying insects for flight
control. Specificities of the medulla and the lobula will be further described
in the following two Sections.

3.3.2 Local Optic-flow Detection

Although the use of optic flow in insects is widely recognised as the primary visual cue for in-flight navigation, the neuronal mechanisms underlying local motion detection in the medulla remain elusive [Franceschini
et al., 1989; Single *et al.*, 1997]. However, behavioural experiments cou-
pled with recordings from the tangential cells in the lobula have led to
functional models of local motion detection. The one best-known is the

so-called *correlation-type elementary motion detector* (EMD), first proposed by Hassenstein and Reichardt [1956], in which intensity changes in neighboring ommatidia are correlated [Reichardt, 1961, 1969]. This model was initially proposed to account for the experimentally observed *optomotor response* in insects [Götz, 1975]. Such a behaviour tends to stabilise the insect's orientation with respect to the environment and is evoked by the apparent movement of the visual environment.

An EMD of the correlation type basically performs a multiplication of input signals received by two neighbouring photoreceptors (Fig. 3.7). Prior to entering the multiplication unit, one of the signals is delayed (e.g. using by a first order low-pass filter), whereas the other remains unaltered. Due to these operations, the output of each multiplication unit preferentially responds to visual stimuli moving in one direction. By connecting two of them with opposite directional sensitivities as excitatory and inhibitory elements to an integrating output stage, one obtains a bidirectional EMD (see also Borst and Egelhaaf, 1989, for a good review of the EMD principle). This popular model has been successful at explaining electrophysiological responses of tangential cells to visual stimuli (see, e.g. Egelhaaf and Borst, 1989) and visually-elicited behavioural responses (see, e.g. Borst, 1990).

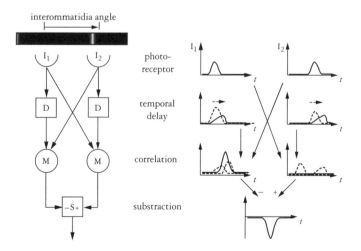

Figure 3.7 The correlation-type elementary motion detector [Reichardt, 1969]. See text for details. Outline adapted from [Neumann, 2003].

On the other hand, it is important to stress that this detector is not a pure image velocity detector. Indeed, it is sensitive to the contrast frequency of visual stimuli and therefore confounds the angular velocity of

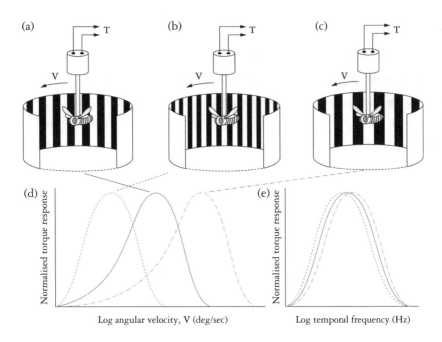

Figure 3.8 The optomotor response of insects [Srinivasan *et al.*, 1999]. If a flying insect is suspended in a rotating striped drum, it will attempt to turn in the direction of the drum's rotation. The resulting yaw torque is a measure of the strength of the optomotor response. For stripes of a given angular period (as in a), the normalised strength of the optomotor response is a bell-shaped function of the drum's rotational speed, peaking at a specific angular velocity of rotation (solid curve, d). If the stripes are made finer (as in b), one obtains a similar bell-shaped curve, but with the peak shifted toward a lower angular velocity (dashed curve, d). For coarser stripes (as in c), the peak response occurs at higher angular velocities (dot-dashed curve, d). However, the normalised response curves coincide with each other if they are re-plotted to show the variation of response strength with the temporal frequency of optical stimulation that the moving striped pattern elicits in the photoreceptors, as illustrated in (e). Thus, the optomotor response that is elicited by moving striped patterns is tuned to temporal frequency rather than to angular velocity. Reprinted with permission from Prof. Mandyam V. Srinivasan.

patterns with their spatial structure [Reichardt, 1969; Egelhaaf and Borst, 1989; Franceschini *et al.*, 1989; Srinivasan *et al.*, 1999][3]. The correlation-type EMDs are tuned to temporal frequency, rather than to angular velocity, as outlined by the summary of the optomotor response experiment in Figure 3.8.

Although visual motion processing in insects has been studied and characterised primarily through the optomotor response, alternative techniques have led researchers to contradictory conclusions with regard to local motion detection. In the 1980's, Franceschini and colleagues proposed a different scheme of local motion detection using lateral facilitation of a high-pass filtered signal [Franceschini *et al.*, 1989; Franceschini, 2004]. This was the result of experiments whereby single photoreceptors of the fly retina were stimulated in sequence while the activity of a specific tangential cell in the lobula was recorded. The underlying idea was that an intensity change detected by a photoreceptor yields a slowly (exponentially) decaying signal that is sampled by an impulse due to the same intensity change when it hits the neighbouring photoreceptor.

Studies with free-flying bees have identified several other visually elicited behaviours that cannot be explained by the optomotor response and the correlation-type EMD model. These behaviours are essentially the centering response, the regulation of flight speed, and the landing strategy (see Section 3.4.4 for further description). All these behaviours appear to be mediated by a motion detection mechanism that is sensitive primarily to the speed of the visual stimulus, regardless of its spatial structure or the contrast frequency that it produces [Srinivasan *et al.*, 1999]. These findings are further supported by an experiment with free-flying Drosophila, where the fruitflies were found to demonstrate a good insensitivity to spatial frequency when keeping ground speed constant by maintaining optic flow at a preferred value, while presented with various upwind intensities [David, 1982].

A neurobiologically realistic scheme for measuring the angular speed of an image, independent of its structure or contrast, has been proposed

---

[3] However, recent work has shown that for natural scenes, enhanced Reichardt EMDs can produce more reliable estimates of image velocity [Dror *et al.*, 2001].

[Srinivasan *et al.*, 1991]. This non-directional model is still hypothetical, although recent physiological studies have highlighted the existence of distinct pathways in the optic lobes responsible for directional and nondirectional motion detection [Douglass and Strausfeld, 1996]. Unlike Reichardt's (correlation-type) and Franceschini's (facilitate-and-sample) models, Srinivasan's model fairly accurately encodes the absolute value of image velocity but not the direction of motion. Note that non-directional motion detection is sufficient for some of the above-mentioned behaviours, such as the centering response.

It is interesting to notice that the Reichardt model is so well established that it has been widely used in bio-inspired robotics (e.g. Huber, 1997; Harrison, 2000; Neumann and Bülthoff, 2002; Reiser and Dickinson, 2003; Iida, 2003). Nevertheless some notable deviations from it exist [Weber *et al.*, 1997; Franz and Chahl, 2002; Ruffier and Franceschini, 2004]. In our case, after preliminary trials with artificial implementation of correlation-type EMDs, it became clear that more accurate image velocity detection (i.e. independent of image contrast and spatial frequency) would be needed for the flying robots. We therefore searched for non-biologically-inspired algorithms producing accurate and directional optic flow estimates. The image interpolation algorithm (also proposed by Srinivasan, see Chapter 5) was selected. To clearly stress the difference, the term optic-flow detector (OFD) is used to refer to the implemented scheme for local motion detection instead of the term EMD. Of course, the fact that local motion detection is required as a preprocessing stage in flying insects is widely accepted among biologists and is thus also applied to the bio-inspired robots presented in this book.

3.3.3   Analysis of Optic-flow Fields

Visual motion stimuli occur when an insect moves in a stationary environment, and their underlying reason is the continual displacement of retinal images during self motion. The resulting optic-flow fields depend in a characteristic way on the trajectory followed by the insect and the 3D structure of the visual surroundings. These motion patterns therefore contain information indicating to the insect its own motion and the distances from potential obstacles. However, this information cannot be directly retrieved at the local level and optic flow from various regions of the visual field must be combined in order to infer behaviourally significant information (Fig. 3.9).

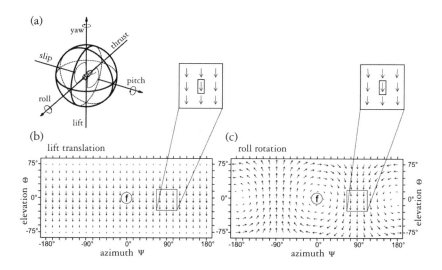

Figure 3.9 The global structures of translational and rotational optic-flow fields. (a) The movements of a fly can be described by their translational (thrust, slip, lift) and rotational (roll, pitch, yaw) components around the 3 body axes (longitudinal, transverse, vertical). These different motion components induce various optic-flow fields over both eyes of the moving insect. For simplicity, equal distances from the objects in a structured environment are assumed. (b) An optic-flow field caused by a lift translation. (c) An optic-flow field caused by a roll rotation. Optic-flow patterns are transformed from the visual unit sphere into Mercator maps to display the entire visual space using spherical coordinates. The visual directions are defined by the angles of azimuth and elevation. The encircled f (frontal) denotes the straight-ahead direction. Globally, the two optic-flow fields can easily be distinguished from one another. However, this distinction is not possible at the level of local motion detectors. See, e.g. the optic-flow vectors indicated in the boxes: local motion detectors at this place would elicit exactly the same response irrespective of the motion. Reprinted from Krapp *et al.* [1998] with permission from The American Physiological Society.

Analysis of the global motion field (or at least several different regions) is thus generally required in order for the local measurements to be exploited at a behavioural level. Some sort of spatial integration is known to take place after the medulla (where local motion detection occurs retinotopically), mainly in the lobula plate where tangential neurons receive input from large receptive fields [Hausen and Egelhaaf, 1989]. The lobula plate thus represents a major centre for optic-flow field analysis. Some of the 60 neurons of the lobula plate are known to be sensitive to coherent large-field motion (i.e. the VS, HS and Hx-cells), whereas other neurons, the Figure detection cells (FD-cells), are sensitive to the relative motion between small objects and the background [Egelhaaf and Borst, 1993b; Krapp and Hengstenberg, 1996]. As an example of the usefulness of these neurons at the behavioural level, there is sound evidence that HS and VS-cells are part of the system that compensates for unintended turns of the fly from its course [Krapp, 2000].

Detection of Self-motion

Quite recently, neuroscientists have analysed the specific organisation of the receptive fields, i.e. the distribution of local preferred directions and local motion sensitivities, of about 30 tangential cells out of the 60 present in the lobula. They found that the response fields of VS neurons resemble rotational optic-flow fields that would be induced by the fly during rotations around various horizontal axes [Krapp et al., 1998]. In contrast to the global rotational structure of VS cells, the response field of Hx cells have the global structure of a translational optic-flow field [Krapp and Hengstenberg, 1996]. The response fields of HS cells are somewhat more difficult to interpret since it is believed that they do not discriminate between rotational and translational components [Krapp, 2000]. In summary, it appears that tangential cells in the lobula act as neuronal matched filters [Wehner, 1987] tuned to particular types of visual wide-field motion (Fig. 3.10). It is also interesting to notice that these receptive-field organisations are highly reliable at the interindividual level [Krapp et al., 1998] and seem to be independent of early sensory experiences of the fly. This suggests that the sensitivity of these cells to optic-flow fields has evolved on a phylogenetic time scale [Karmeier et al., 2001].

Franz and Krapp [2000] experienced a certain success when estimating self-motion of a simulated agent based on this theory of visual matched filters. However, Krapp [2000] interprets this model of spatial integration with caution:

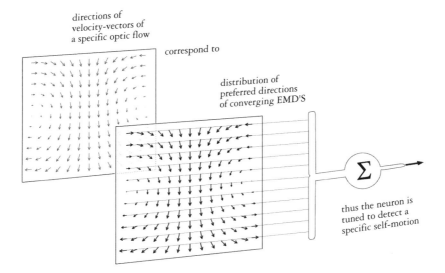

Figure 3.10 A hypothetical filter neuron matched to a particular optic-flow field induced by self-motion (e.g. rotation). Local motions of the optic-flow field locally activate those motion detectors with appropriate preferred directions. A wide-field neuron selectively collects and spatially integrates the signals of these motion detectors. Hence, it would be most sensitive to that particular optic-flow and consequently to the self-motion that caused the flow. Reprinted from Krapp *et al.* [1998] with permission from The American Physiological Society.

> [Some] approaches take for granted that the results of the local motion estimates are summed up in a linear fashion at an integrating processing stage. For insect visual systems, however, it was found that local motion analysis is achieved by elementary motion detectors whose output is not simply proportional to velocity (...) but also depends on pattern properties like spatial wavelength and contrast (...). Hence, it remains unclear how biological sensory systems cope with highly dynamic stimuli as encountered, for instance, by the fly during free flight. It is by no means easy to predict the signals of the tangential neurons under such natural conditions.

Another problem is that tangential neurons, such as the VS cells, cannot be expected to be insensitive to optic-flow components induced by movements that are not their own preferred self-motion. The output from those neurons needs to be corrected for apparent rotations, which may be due to translational self motions and to rotations around axes other than the preferred axis. In fact, the use of visual or gyroscopic information for correcting such errors is a recurrent question which has yet to be resolved. According to Krapp [2000],

> The signals necessary to correct for these erroneous response contributions could be supplied by other wide field neurons.

Or, alternatively:

> Correction signals encoding fast self-rotations may also be supplied by the haltere system [Nalbach, 1994]. Because the dynamic range of the haltere system is shifted toward higher angular velocities, it is thought to complement the visual self-motion estimation [Hengstenberg, 1991].

The computational properties of tangential neurons have mainly been characterised in tethered flies with simplistic visual stimuli (e.g. Krapp *et al.*, 1998). A recent study where blowflies were presented with behaviourally relevant visual inputs suggests that responses of tangential cells are very complex and hard to predict based on the results obtained with simplistic stimuli [Lindemann *et al.*, 2003]. As explained by Egelhaaf and Kern [2002], only few experiments with natural stimuli have been performed and even less in closed-loop situation:

> Neuronal responses to complex optic flow as experienced during unrestrained locomotion can be understood only partly in terms of the concepts that were established on the basis of experiments done with conventional motion stimuli. (...) It is difficult to predict the performance of the system during complex flight manoeuvres, even when wiring diagrams and responses to simplified optic-flow stimuli are well established.

Perception of Approaching Objects

Apart from the widely covered topic of tangential cells in the lobula plate and their resemblance to matched filters, another model of wide field integration has been proposed to explain the detection of imminent collision. Here, the purpose is to estimate the distance from objects or the *time to contact* (TTC), rather than to detect self motion. Looming stimuli (expanding images) have long been thought to act as essential visual cues for detecting imminent collisions (see, e.g. Lee, 1976). When tethered flying flies encounter a looming stimulus, they extend their forelegs in preparation for landing. This *landing response* has been shown to be triggered by visual looming cues [Borst and Bahde, 1988]. Experiments demonstrate that the latency of the landing response is reciprocally dependent on the spatial frequency content and on the contrast of the pattern, as well as on the duration of its expansion. Borst and colleagues have proposed a model based on a spatial integration of correlation-type EMDs (Fig. 3.11), which presents the same kind of dependence on spatial frequency and contrast (see Sect. 3.3.2). Very recently, Tammero and Dickinson [2002b] have shown that collision avoidance manoeuvres in fruitflies can also be explained by the perception of image expansion as detected by an array of local motion detectors (see Sect. 3.4.3).

So far, neurons that extract image expansion from the retinotopic array of local motion detectors have not been found at the level of the lobula complex [Egelhaaf and Borst, 1993b]. In the cervical connective (just below the brain in Figure 3.6), however, cells are known to be sensitive to retinal image expansion. These neurons, which respond strongly when the insect approaches an obstacle or a potential landing site, have been proposed to be part of the neuronal circuit initiating the landing response [Borst, 1990].

Other biologists have proposed similar schemes, although based on pure TTC and thus without any dependency on contrast or spatial frequency, for explaining the deceleration of flies before landing [Wagner, 1982] or the stretching of their wings in plunging gannets [Lee and Reddish, 1981]. From a functional point of view, it would obviously be advantageous to use a strategy that estimates TTC independently of the spatial structure of the object being approached. Indeed, if the underlying local optic-flow detection is a true image velocity detection, the measure of the

TTC can be directly extracted from optic-flow measurements [Poggio *et al.*, 1991; Ancona and Poggio, 1993; Camus, 1995].

In summary, individual cells (either in the lobula or in the cervical connective) receive inputs from many local motion detectors and generate output signals that appear to be tuned to estimate particular features of the global optic-flow field that flying insects experience during flight. Spatial integration of local optic-flow vectors is thus a necessary operation to provide useful information for several behaviours such as stabilisation, landing, collision avoidance, etc. Although the weight limitations of the flying platforms do not permit the presence of as many local motion detectors as in flying insects, some kind of spatial integration (e.g. combining signals from left and right OFDs) is used to detect typical patterns of optic flow.

3.4 In-Flight Behaviours

As previously described, flying insects use visual motion and mechanosensors to gain information on the 3D layout of the environment and the rate of self-motion in order to control their behaviours. In this Section, a set of basic behaviours is reviewed and linked to possible underlying information processing strategies presented in the previous Section. This restricted palette of behaviours is not a representative sample of the biological literature, but rather a minimal set of control mechanisms that would allow a flying system to remain airborne in a confined environment.

3.4.1 Attitude Control

One of the primary requirements for a flying system is to be able to control its *attitude* in order to stay upright or bank by the right amount during turns [Horridge, 1997]. The attitude of an aerial system is defined by its pitch and roll angles (Fig. 3.9a). The so-called *passive stability* encompasses simple mechanisms providing flight stability without active control. For instance, the fact that insect wings are inserted above the center of gravity provides some degree of passive stability around the roll axis [Chapman, 1998, p. 214]. Other aerodynamic characteristics of the insect body provide partial compensation for unintended pitch torques [Dudley, 2000, p. 203].

However, in small flapping-wing insects relying on unsteady-state aerodynamics[4], such passive mechanisms can compensate only for a small subset of unintentional rotations.

Insects thus require other mechanisms for attitude control. One such mechanism is the so-called *dorsal light response* [Schuppe and Hengstenberg, 1993] by which insects attempt to balance the level of light received in each of their three ocelli (see Sect. 3.2.1). This response is believed to help insects keep their attitude aligned with the horizon [Dudley, 2000, p. 212]. Such mechanisms have been proposed for attitude control in simulated flying agents [Neumann and Bülthoff, 2002]. However, this approach is not viable in indoor environments, since there exits no horizon nor a well defined vertical light gradient. If insects could control their attitude exclusively by means of a dorsal light response, they would demonstrate a tendency to fly at unusual angles when flying among obstacles that partially occlude light sources. The fact that this does not occur indicates the importance of other stimuli, although they are not yet fully understood [Chapman, 1998, p. 216].

It is probable that optic flow (see Sect. 3.3.3) provides efficient cues for pitch and roll stabilisation in a functionally similar manner to the optomotor response (primarily studied for rotations around the yaw axis). However, optic flow depends on the angular rate and not on absolute angles. Angular rates must be integrated over time to produce absolute angles, but integration of noisy rate sensors results in significant drift over time. Therefore, such mechanisms fail to provide reliable information with respect to the attitude. The same holds true for the halteres (see Sect. 3.2.2), which are also known to help at regulating pitch and roll velocities but are not able to provide an absolute reference over long periods of time.

In artificial systems, such as aircraft relying on steady-state aerodynamics, passive stabilisation mechanisms are often sufficient in providing compensation torques to progressively eliminate unintended pitch and roll. For instance, a positive angle between the left and right wings (called dihedral, see Section 4.1.3 for further details) helps in maintaining the wings horizontal, whereas a low center of gravity and/or a well-studied tail geometry

---

[4] Direction, geometry and velocity of airflow change over short time intervals.

provides good pitch stability[5]. The aircraft described later on in this book operate within the range of steady-state aerodynamics and therefore do not need an active attitude control, such as the dorsal light response.

3.4.2 Course (and Gaze) Stabilisation

Maintaining a stable flight trajectory is not only useful when travelling from one point to another, but it also facilitates depth perception, as pointed out by Krapp [2000]:

> Rotatory self-motion components are inevitable consequences of locomotion. The resulting optic-flow component, however, does not contain any information about the 3D layout of the environment. This information is only present within translational optic-flow fields. Thus for all kinds of long-range and short-range distance estimation tasks, a pure translatory optic-flow field is desirable [Srinivasan et al., 1996, (...)]. One possibility to, at least, reduce the rotatory component in the optic-flow is to compensate for it by means of stabilising head movements and steering manoeuvres. These measures can be observed in the fly but also in other visually oriented animals, including humans.

The well-known optomotor response (introduced in Section 3.3.2), which is evoked by the apparent movement of the visual environment, tends to minimise image rotation during flight and helps the insect to maintain a straight course [Srinivasan et al., 1999]. Hence, course stabilisation of flying insects relies essentially on the evaluation of the optic-flow patterns perceived during flight and reviewed in Section 3.3.3. Haltere feedback is also known to play an important role in course stabilisation as well as in gaze or head[6] orientation. As suggested in Krapp's statement, a rapid head compensation aids in cancelling rotational optic-flow before the rest of the body has time to react (see also Hengstenberg, 1991). For instance, in the

---

[5] Note however, that rotorcrafts are far less passively stable than airplanes and active attitude control is a delicate issue because proprioceptive sensors like inclinometers are perturbed by centripetal accelerations during manoeuvres.

[6] In this context, gaze and head control have the same meaning as a result of insect eyes being mostly solidly attached to the head.

free-flying blowfly the angular velocities of the head are approximately half those of the thorax during straight flight [van Hateren and Schilstra, 1999].

The integration of visual and gyroscopic senses for course and gaze stabilisation in flying insects seems intricate and is not yet fully understood. Chan *et al.* [1998] have shown that motoneurons innervating the muscles of the haltere receive strong excitatory input from visual interneurons such that visually guided flight manoeuvres may be mediated in part by efferent modulation of hard-wired equilibrium reflexes. Sherman and Dickinson [2004] have proposed a stabilisation model where sensory inputs from the halteres and the visual system are combined in a weighted sum. What is better understood, though, is that fast rotations are predominantly detected and controlled by mechanosensory systems whereas slow drifts and steady misalignments are perceived visually [Hengstenberg, 1991].

Whatever the sensory modality used to implement it, course stabilisation is clearly an important mechanism in visually guided flying systems. On the one hand, it enables counteractions to unwanted deviations due to turbulences. On the other hand, it provides the visual system with less intricate optic-flow fields (i.e. exempt of rotational components), hence facilitating depth perception, and eventually collision avoidance.

3.4.3 Collision Avoidance

As seen in Section 3.3.3, a trajectory aiming at a textured object or surface would generate strong looming cues, which can serve as imminent collision warnings. Various authors have shown that the deceleration and extension of the legs in preparation for landing are triggered by large-field, movement-detecting mechanisms that sense an expansion of the image [Borst and Bahde, 1988; Wagner, 1982; Fernandez Perez de Talens and Ferretti, 1975]. Instead of extending their legs for landing, flying insects could decide to turn away from the looming object in order to avoid it.

This subject has been recently studied by Tammero and Dickinson [2002a]. The flight trajectories of many fly species consist of straight flight sequences[7] interspersed with rapid changes in heading known as *saccades*

---

[7] During which the course stabilisation mechanisms described above are probably in action.

[Collett and Land, 1975; Wagner, 1986; Schilstra and van Hateren, 1999]. Tammero and Dickinson [2002a] have reconstructed the optic flow seen by free-flying Drosophila. Based on the recorded data, they proposed a model of saccade initiation using the detection of visual expansion, a hypothesis that is consistent with the open-loop presentation of expanding stimuli to tethered flies [Borst, 1990]. Although differences in the latency of the collision-avoidance reaction with respect to the landing response suggest that the two behaviours are mediated by separate neuronal pathways [Tammero and Dickinson, 2002b], the STIM model proposed by Borst [1990] and reprinted in Figure 3.11 represents a good understanding of the underlying principle. Several implementations of artificial systems capable of avoiding collisions have been carried out using a variant of this model. The artificial implementation that was the most closely inspired by the experiments of Tammero and Dickinson [2002a] was developed in the same laboratory (Reiser and Dickinson, 2003, see also Section 2.2).

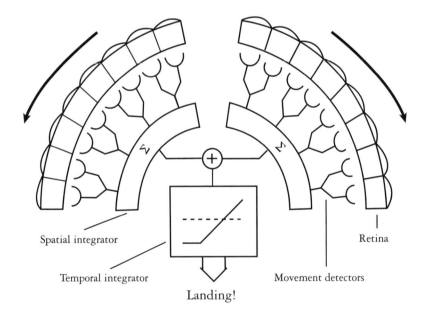

Spatial integrator

Temporal integrator

Retina

Movement detectors

Landing!

Figure 3.11 The so-called STIM (spatio-temporal integration of motion) model underlying the landing response of the fly [Borst and Bahde, 1988]. The output of directionally selective correlation-type movement detectors are pooled from each eye. These large-field units feed into a temporal leaky integrator. Whenever the integrated signal reaches a fixed threshold, landing is released and a preprogrammed leg motor sequence is performed to avoid crash-landing.

3.4.4   Altitude Control

Altitude control is a mechanism that has rarely been directly studied in insects. However, it obviously represents an important mechanism for roboticists with respect to the building of autonomous flying machines. In this Section, we thus consider related behaviours in flying insects that help to understand how an aerial system could regulate its altitude by using visual motion cues. These behaviours – especially studied in honeybees – are the centering response, the regulation of the flight speed, and the grazing landing.

Bees flying through a narrow gap or tunnel have been shown to maintain equidistance to the flanking walls (centering response) by balancing the apparent speeds of the retinal images on either side [Srinivasan et al., 1996, 1997]. The experiments reported by Srinivasan et al. [1991] unequivocally demonstrate that flying bees estimate lateral distances from surfaces in terms of apparent motion of their images irrespective of their spatial frequency or contrast.

In another set of experiments [Srinivasan et al., 1996, 1997; Srinivasan, 2000], the speed of flying bees has been shown to be controlled by maintaining a constant optic flow in the lateral regions of the two eyes. This arguably avoids potential collisions by ensuring that the insect slows down when flying through narrow passages.

The grazing landing (as opposed to the landing response described in Section 3.4.3) describes how bees execute a smooth touchdown on horizontal surfaces [Srinivasan et al., 1997, 2000]. In this situation, looming cues are weak as a result of the landing surface being almost parallel to the flight direction. In this case, bees have been shown to hold the image velocity of the surface in the ventral part of their eyes constant as they approach it, thus automatically ensuring that the flight speed is close to zero at touchdown.

These three behaviours clearly demonstrate the ability of flying insects to regulate self-motion using translational optic-flow. The advantage of such strategies is that the control is achieved by a very simple process that does not require explicit knowledge of the distance from the surfaces [Srinivasan and Zhang, 2000].

Observation of migrating locusts have shown that these animals tend to maintain the optic flow experienced in the ventral part of their eyes constant [Kennedy, 1951]. This ventral optic flow is proportional to the ratio between forward speed and altitude. Taking inspiration from these observations, Ruffier and Franceschini [2003] proposed an altitude control system, an optic-flow regulator, that keeps the ventral optic flow at a reference value. At a given ground speed, maintaining the ventral optic flow constant leads to level flight at a given height. If the forward speed happens to decrease (deliberately or as a consequence of wind), the optic flow regulator produces a decrease in altitude. This optic-flow regulator was implemented on a tethered helicopter and demonstrated an efficient altitude control and terrain following. Ruffier and Franceschini [2004] also showed that the same strategy could generate automatic takeoff and landing, and suitable descent or ascent in the presence of wind [Franceschini et al., 2007], as actually observed in migrating locusts [Kennedy, 1951].

One of the major problems of such strategies lies, once again, in the perturbation of the translational flow field by rotational components. In particular, every attitude correction will result in rotation around the pitch or roll axes and indeed create a rotational optic flow. Hence, a system correcting for these spurious signals is required. In flying insects, this seems to be the role of gaze stabilisation (described in Section 3.4.2). In artificial systems, the vision system could be actively controlled so as to remain vertical (this solution was adopted in Ruffier and Franceschini, 2004). However, such a mechanism requires a means of measuring attitude angles in a non-inertial frame, which is a non-trivial task. Another solution consists of measuring angular rates with an inertial system (rate gyro) and directly subtracting rotational components from the global optic-flow field (derotation).

3.5   Conclusion

Attitude control (see Sect. 3.4.1) in insects is believed to be required in order to provide a stable reference for using vision during motion [Horridge, 1997]; and in turn, vision seems to be the primary cue for controlling attitude. The same holds true for course stabilisation (see Sect. 3.4.2), whereby straight trajectories allow for the cancellation of rotational optic flow and an easier interpretation of optic flow for distance estimation. This shows, once again, that perception, information processing, and behaviour are tightly interconnected and organised into a loop where adequate behaviour is not only needed for navigation (and, more generally, survival), but also represents a prerequisite for an efficient perception and information processing. This idea is equally highlighted by biologists like Egelhaaf *et al.* [2002]:

> Evolution has shaped the fly nervous system to solve efficiently and parsimoniously those computational tasks that are relevant to the survival of the species. In this way animals with even tiny brains are often capable of performing extraordinarily well in specific behavioural contexts.

Therefore, when taking inspiration from biology, it is worth perceiving these different levels as tightly connected to each other, rather than trying to design artificial systems behaving like animals while featuring highly precise, Cartesian sensors, or, contrarily, creating robots with biomorphic sensors for cognitive tasks. Following this trend, our approach to robot design will take inspiration from flying insects at the following three levels:

- *Perception.* The choice of sensor modalities is largely based on those of flying insects (Chap. 4). Only low-resolution vision, gyroscopic and airflow information will be fed to the control system.

- *Information processing.* In the experiments described in Chapter 6, the manner of processing information is largely inspired by what has been described above. Visual input is first preprocessed with an algorithm producing local optic-flow estimates (Chap. 5), which are then spatially integrated and combined with gyroscopic information in order to provide the control system with meaningful information.

- *Behaviour.* Based on this preprocessed information, the control system is then designed to loosely reproduce the insect behaviours presented in Section 3.4, which are tuned to the choice of sensors and processing. The resulting system will provide the robots with the basic navigational capability of moving around autonomously while avoiding collisions.

CHAPTER 4

Robotic Platforms

As natural selection is inherently opportunistic, the neurobiologist must adopt the attitude of the engineer, who is concerned not so much with analyzing the world than with designing a system that fulfils a particular purpose.

R. Wehner, 1987

This Chapter presents mobile robots that have been specifically developed to assess bio-inspired flight control strategies in real-world conditions. These include a miniature wheeled robot for preliminary tests, an indoor airship, and two ultra-light fixed-wing airplanes. In spite of the fundamental differences regarding their body shapes, actuators and dynamics, the four robotic platforms use several of the same electronic components, such as sensors and processors, in order to ease the transfer of software, processing schemes and control strategies from one to the other. Obviously, these robots do not attempt to reproduce the bio-mechanical principles of insect flight. However, the perceptive modalities present in flying insects are taken into account in the selection of sensors. After presenting the platforms, we will also briefly describe the software tools used to interface with the robots and to simulate them. This Chapter is concluded with an overview of the test arenas and their respective characteristics.

4.1   Platforms

The robotic platforms are introduced in order of increasing complexity of their dynamic behaviour. This Section focuses on the mechanical architecture and the dynamic behaviour of the different robots, whereas the next Section presents their electronic components and sensors, which are largely compatible among the three platforms. At the end of the Section, a comparative summary of the main characteristics of the platforms is provided.

4.1.1 Miniature Wheeled Robot

The popular *Khepera* [Mondada *et al.*, 1993] was defined as our battle horse for preliminary testing of control strategies. The *Khepera* is a simple and robust differential-drive robot that has proven suitable for long-lasting experiments that are typical in evolutionary robotics (see Sect. 7.3.1). It can withstand collisions with obstacles, does not overheat when its motors are blocked, and can be powered externally via a rotating contact hanging above the test arena, thereby relieving the experimenter of the burden of constantly changing batteries.

To enable a good compatibility with the following aerial platforms, the *Khepera* is augmented with a custom turret (Fig. 4.1). The so-called *kevopic* (*Khepera*, evolution, PIC) turret features the same small microcontroller and interfacing capabilities as the boards mounted on the flying robots. The *kevopic* also supports the same vision and gyroscopic sensors as the one equipping the flying robots (see Sect. 4.2.2).

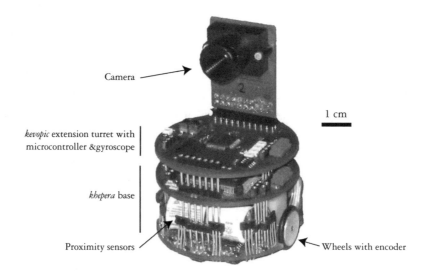

Camera

1 cm

kevopic extension turret with microcontroller &gyroscope

khepera base

Proximity sensors

Wheels with encoder

Figure 4.1 The *Khepera* robot equipped with the custom extension turret *kevopic*.

The sensing capabilities of the underlying standard *Khepera* remain accessible from the custom-developed *kevopic*. Besides the two main sensor modalities (vision and gyroscope) attached to the *kevopic*, the *Khepera* base features 2 wheel encoders and 8 infrared proximity sensors. These additional sensors are useful for analysing the performances of the bio-inspired controllers. For instance, the proximity sensors can be used to detect whether the robot is close to the arena boundaries and the wheel encoders enable the plotting of the produced trajectories with a reasonable precision over a relatively short period of time.

The *Khepera* moves on a flat surface and has 3 degrees of freedom (DOF). It is therefore an ideal candidate for testing collision avoidance algorithms without the requirement of course stabilisation. Since it is in contact with the floor and has negligible inertial forces, the trajectory is determined solely by the wheel speeds. It suffices to issue the same motor command on the left and on the right wheels to obtain a straight trajectory. Of course, attitude and altitude control are not required on this robot. However, Chapter 6 describes how the *Khepera* is employed to demonstrate vision-based altitude control by orienting the camera laterally and performing wall following. From a bird-eye perspective, the wall replaces the ground and, at a first approximation, the heading direction of the *Khepera* is similar to the pitch angle of an airplane.

4.1.2 Blimp

When it comes to flying robots, one has to choose a method of producing lift among those existing: aerostat, fixed-wing, flapping-wing, rotorcraft, and jet-based. The simplest method from both a mechanical and structural point of view is the aerostat principle.

Blimps as Robotic Platforms

According to Archimedes, a volume surrounded by a fluid (in our case, the ambient air) generates a buoyant force that is equal to the mass of the fluid displaced by this volume. In order to fly, airships must thus be lighter than the mass of the air occupied by their hull. This achieved by filling the volume of their hull with a gas far lighter than air (helium is often employed) in order to compensate for the weight of the gondola and

equipment that is hanging below the hull. Such a lift principle presents
several advantages:

- No specific skills in aerodynamics are needed for building a system
  able to fly. Inflating a bag with helium and releasing it into the air
  with some balancing weight produces a minimalist flying platform that
  remains airborne in much the way that a submarine stays afloat in water.

- Unlike helicopters or jet-based systems, it is not dangerous for indoor
  use and is far quieter.

- Unlike all other flying schemes, it does not require energy to stay aloft.

- The envelope size can easily be adapted to the required payload (e.g. a
  typical spherical Mylar bag of 1 m in diameter filled with helium can
  approximately lift 150 g of payload in addition to its own weight).

- An airship is stable by nature. Its center of gravity lies below the center of buoyancy, creating restoring forces that keep the airship upright.
  If used under reasonable accelerations, an airship can thus be approximated by a 4 DOF model because pitch and roll angles are always close
  to zero.

- Equipped with a simple protection, a blimp can bump into obstacles
  without being damaged while remaining airborne, which is definitely
  less than trivial for airplanes or helicopters.

All these advantages have led several research teams to adopt such lighter-
than-air platforms in various areas of indoor robotic control such as visual servoing [Zhang and Ostrowski, 1998; van der Zwaan *et al.*, 2002;
da Silva Metelo and Garcia Campos, 2003], collective intelligence
[Melhuish and Welsby, 2002], or bio-inspired navigation [Planta *et al.*,
2002; Iida, 2003]. The same advantages allowed us to set up the first evolutionary experiment entirely performed on a physical flying robot [Zufferey
et al., 2002]. Note that the version used at that time, the so-called *Blimp1*,
was slightly different from the one presented here.

Apart from the need for periodic refills of the envelope, the main drawbacks of a blimp-like platform reside in its inertia due to its considerable
volume. Because of its shape and dynamics, a blimp also has less in common with flying insects than an airplane. This platform was mainly built
as an intermediate step between the miniature wheeled robot and the ultra-

light winged airplanes to enable aerial experiments that would not be possible with airplanes (Chap. 7). Although a blimp is probably the simplest example of a platform capable of manoeuvring in 3D, it already has much more complex dynamics than a small wheeled robot because its inertia and tendency to side slip.

The *Blimp2b*

The most recent prototype, the so-called *Blimp2b* (Fig. 4.2), has a helium-filled envelope with a lift capacity of 100 g. The near-ellipsoid hull measures $110 \times 60 \times 60$ cm. The gondola underneath consists of thin carbon rods. Attached to the gondola frame are three thrusters (8-mm DC motors, gears and propellers from Didel SA[1]), a horizontal 1D camera pointed forward, a yaw rate gyro, an anemometer and a distance sensor (Sharp™ GP2Y0A02YK) measuring the altitude above the ground. The on-board energy is supplied by a 1200 mAh lithium-polymer battery, which is sufficient for 2-3 hours of autonomy.

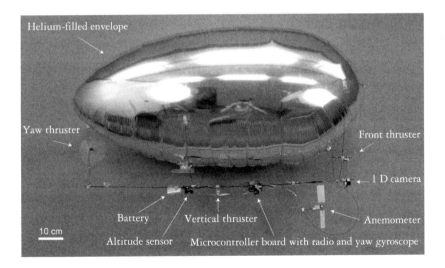

Figure 4.2 The autonomous indoor airship *Blimp2b* with a description of all its electronic components, sensors and actuators.

---

[1] http://www.didel.com

Although the *Blimp2b* can move in 3D, roll and pitch movements
are passively stabilised around the horizontal attitude. Consequently, the
Blimp2b has virtually only 4 DOF. Furthermore, an automatic altitude control using the vertical distance sensor can be enabled to reduce the manoeuvring space to 2D and the number of DOF to 3 instead of 4. Even with
this simplification, the airship displays much more complex dynamics with
respect to the *Khepera* and, furthermore, no trivial relation exists between
the voltages applied to the motors and the resulting trajectory. This is due
to inertia (not only of the blimp itself but also of the displaced air in the
surroundings of the hull) and to aerodynamic forces [Zufferey *et al.*, 2006].
Therefore, in addition to collision avoidance, the *Blimp2b* requires course
stabilisation in order to move forward without rotating randomly around its
yaw axis. On the other hand, vision-based altitude control is not required
when using the vertical distance sensor, and the natural passive stabilisation
means that an active attitude control is also not necessary.

4.1.3   Indoor Airplanes

In 2001, together with the EPFL spin-off Didel SA, the process of developing ultra-light flying airplanes for indoor robotic research was started
[Nicoud and Zufferey, 2002]. Rotorcrafts and flapping-wing systems (see
Section 2.1 for a review) were discarded mainly because of their mechanical complexity, their intrinsic instability and the lack of literature concerning unsteady-state aerodynamics at small scales and low speed (i.e. low
Reynolds number). Instead, efforts were aimed at a simple platform capable
of flying in office-like environments; a task that requires a relatively small
size, high manoeuvrability and low-speed flight.

Requirements for Indoor Flying

To better appreciate the challenges of indoor flying, let us review some
basics of steady-state aerodynamics. First of all, the lift F_L and drag F_D
forces acting on a wing of surface S going through the air at velocity v are
given by:

$$F_{L,D} = \frac{1}{2} \rho v^2 S C_{L,D} \,, \tag{4.1}$$

where ρ is the air density and C_L and C_D the lift and drag coefficients, respectively. These coefficients depend on the airfoil geometry, its angle of attack and the airflow characteristics surrounding it. The airflow's (or any fluid's) dynamic characteristics are represented by the dimensionless Reynolds number Re, which is defined as:

$$\text{Re} = \frac{\rho v L}{\mu} = \frac{\rho v^2}{\frac{\mu v}{L}} = \frac{\text{inertial forces}}{\text{viscous forces}}, \qquad (4.2)$$

where μ is the air dynamic viscosity and L a characteristic length of the airfoil (generally the average wing chord, i.e. the distance from leading edge to trailing edge). Re provides a criterion for dynamic similarity of airflows. In other words, two objects of identical shapes are surrounded by similar fluid flows if Re is the same, even if the scales or the type of fluids are different. If the fluid density and viscosity are constant, the Reynolds number is mainly a function of airspeed v and wing size L. The Reynolds number is essentially the relative significance of the viscous effect compared to the inertial effect. Obviously, Re is small for slow-flying, small aerial devices (typically $0.3\text{-}5 \cdot 10^3$ in flying insects, $1\text{-}3 \cdot 10^4$ in indoor slow-flyers), whereas it is large for standard airplanes flying at high speed (10^7 for a Cessna, up to 10^8 for a Boeing 747). Therefore, very different airflows are expected between a small and slow flyer and a standard aircraft. In particular, viscous effects are predominant at small size.

The aerodynamic efficiency of an airfoil is defined in terms of its maximum lift-to-drag ratio [Mueller and DeLaurier, 2001]. Unfortunately, this ratio has a general tendency to drop quickly as the Reynolds number decreases (Fig. 4.3). In addition to flying at a regime of bad aerodynamic efficiency (i.e. low C_L and high C_D), indoor flying platforms are required to fly at very low speed (typically 1-2 m/s), thus further reducing the available lift force F_L produced by the wing (equation 4.1). For a given payload, the only way of satisfying such constraints is to have a very low wing-loading (weight to wing surface ratio), which can be achieved by widening the wing surface without proportionally increasing the weight of the structure. Figure 4.4 shows the place of exception occupied by indoor flying robots among other aircraft. It also highlights the fundamental difference between indoor airplanes and outdoor MAVs [Mueller, 2001]. Although their overall weight

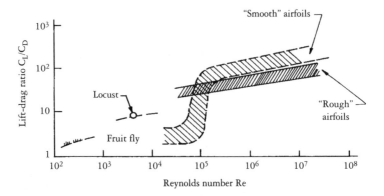

Figure 4.3 The maximum lift-to-drag ratio. The airfoil performance deteriorates rapidly as the Reynolds number decreases below 10^5. Reprinted from McMasters and Henderson [1980] with permission from of the Journal of Technical Soaring and OSTIV.

is similar, their respective speed ranges are located on opposite sides of the trend line. As opposed to indoor flying robots, MAVs tend to have small wings (around 15 cm, to ease transport and pre-launch handling), and fly at high speed (about 15 m/s).

Because of the lack of methods for designing efficient airframe geometries at Reynolds numbers below $2 \cdot 10^5$ [Mueller and DeLaurier, 2001], we proceeded by trial and error. Note that despite the availability of methods for analytical optimisation of airfoils, it would have been exceedingly difficult, if not impossible, to guarantee the shape of the airfoil because of the use of ultra lightweight materials. Moreover, the structural parts of such lightweight airframes are so thin that it is impossible to assume that they do not deform in flight. This may results in large discrepancies between the theoretical and actual airframe geometries and therefore invalidate any *a priori* calculations. Our approach is thus to first concentrate on what can be reasonably built (materials, mechanical design) to satisfy the weight budget and subsequently improve the design on the basis of flight tests and wind tunnel experiments.

Our indoor airplanes are made of carbon-fiber rods and balsa wood for the structural part, and of a thin plastic film (2.2 g/m^2) for the lifting surfaces. Wind tunnel tests allowed the optimisation of the wing structure and airfoil by measuring lift and drag for different wing geometries

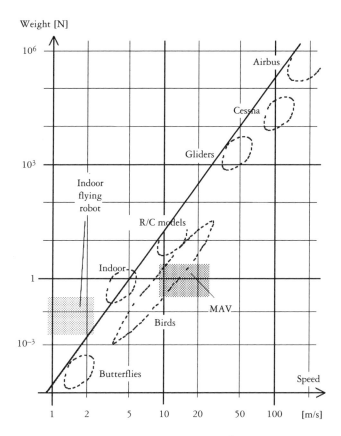

Figure 4.4 Typical aircraft weight versus speed [Nicoud and Zufferey, 2002]. "R/C models" denote typical outdoor radio-controlled airplanes. "Indoor" represents the models used by hobbyists for flying in gymnasiums. These have less efficiency constraints than "Indoor flying robots" since they can fly faster in larger environments. "MAV" stands for micro air vehicles (as defined by DARPA).

[Zufferey *et al.*, 2001]. The measurements were obtained by using a custom-developed aerodynamic scale capable of detecting very weak forces and torques. Furthermore, by employing visualisation techniques (Fig. 4.5a), we were able to analyse suboptimal airflow conditions and modify the airframe accordingly.

Since 2001, various prototypes have been developed and tested. The first operational one was the *C4* (Fig. 4.5b). Weighing 47 g without any sensors (see Zufferey *et al.*, 2001, for the weight budget), this 80 cm-wingspanned airplane was able to fly between 1.4 and 3 m/s with a turning

radius of approximately 2 m. The NiMh batteries used at that time provided an autonomy of a mere 5 minutes.

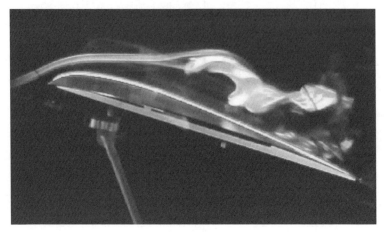

(a) Airflow visualisation

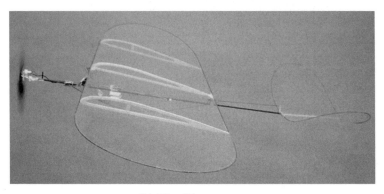

(b) The *C4* prototype

Figure 4.5 (a) Airflow visualisation over the airfoil of the *C4* prototype using a
smoke-laser technique within a special wind tunnel at low air speed. The prototype
is attached to the top of a custom-developed device for measuring very small lift
and drag forces. (b) Preliminary prototype (*C4*) of our indoor airplane series.

The *F2* Indoor Flyer

A more recent version of our robotic indoor flyers, the *F2* (Fig. 4.6), has a
wingspan of 86 cm and an overall weight of 30 g including two vision sensors and a yaw rate gyro (Table 4.1). Thanks to its very low inertia, the *F2*
rarely becomes damaged when crashing into obstacles. This characteristic

is particularly appreciated during early phases of control development. In order to further limit the risk of damaging the aircraft, the walls of the test arena used for this robot are made of fabric (Sect. 4.4).

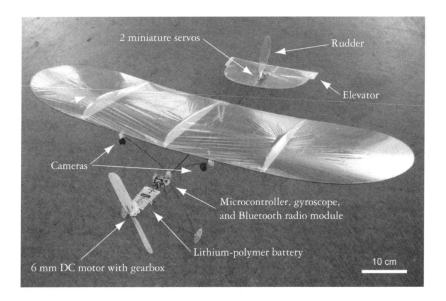

Figure 4.6 The *F2* indoor slow-flyer. The on-board electronics consist of a 6 mm geared motor with a balsa-wood propeller, two miniature servos controlling the rudder and the elevator, a microcontroller board with a Bluetooth module and a rate gyro, two horizontal 1D cameras located on the leading edge of the wing, and a 310 mAh lithium-polymer battery.

Table 4.1 Mass budget of the *F2* prototype.

Subsystem	Mass [g]
Airframe	10.7
Motor, gear, propeller	2.7
2 servos	2.7
Lithium-polymer battery	6.9
Microcontroller board with gyro	3.0
Bluetooth radio module	1.0
2 cameras	2.0
Bluetooth radio module	1.0
Total	30

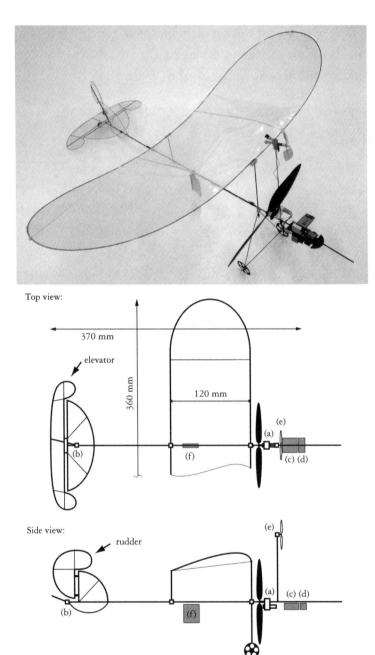

Figure 4.7 The 10-gram *MC2* microflyer. The on-board electronics consists of (a) a 4 mm geared motor with a lightweight carbon fiber propeller, (b) two magnet-in-a-coil actuators controlling the rudder and the elevator, (c) a microcontroller board with a Bluetooth module and a ventral camera with its pitch rate gyro, (d) a front camera with its yaw rate gyro, (e) an anemometer, and (f) a 65 mAh lithium-polymer battery.

The *F2* flight speed lies between 1.2 and 2.5 m/s and its yaw angular rate is in the ±100°/s range. At 2 m/s, the minimum turning radius is less than 1.3 m. The *F2* is propelled by a 6 mm DC motor with a gearbox driving a balsa-wood propeller. Two miniature servos (GD-servo from Didel SA) are placed at the back end of the fuselage to control the rudder and elevator. The on-board energy is provided by a 310 mAh lithium-polymer battery. The power consumption of the electronics (including wireless communication) is about 300 mW, and the overall peak consumption, including the motors, reaches 2 W. The in-flight energetic autonomy is around 30 minutes.

In order to provide this airplane with a sufficient passive stability around roll and pitch angles, the wing was positioned rather high above the fuselage and the tail was located relatively far behind the wing. In addition, a certain dihedral[2] naturally appears in flight because of the distortion of the longitudinal carbon rods holding the wings. This dihedral contributes to the passive roll stability. As a results, no active attitude control is actually needed in order for the *F2* to stay upright in flight. However, course stabilisation can still be useful to counteract air turbulences and the effects of airframe asymmetries. Collision avoidance remains the central issue when automating such an airplane and an altitude controller would also be required. However, this will not be demonstrated on this prototype, but on its successor.

The *MC2* Indoor Microflyer

The latest prototype of our indoor flyers is the *MC2* (Fig. 4.7). This flyer is based on a remote-controlled 5.2-gram home flyer produced by Didel SA for the hobbyist market, and the model consists mainly of carbon fiber rods and thin Mylar plastic films as does the *F2*. The wing and the battery are connected to the frame by small magnets such that they can easily be taken apart. Propulsion is produced by a 4-mm brushed DC motor, which trans-

---

[2] A dihedral is the upward angle of an aircraft's wings from root to tip, as viewed from the front of an aircraft. The purpose of the dihedral is to confer stability in the roll axis. When an aircraft with a certain dihedral is yawing to the left, the dihedral causes the left wing to experience a greater angle of attack, which increases lift. This increased lift tends to cause the aircraft to then return to level flight.

mits its torque to a lightweight carbon-fiber propeller via a 1 : 12 gearbox. The rudder and elevator are actuated by two magnet-in-a-coil actuators. These extremely lightweight actuators are not controlled in position like conventional servos, but, because they are driven by bidirectional pulse width modulated (PWM) signals, they are proportional in torque.

The stock model airplane has been transformed into a robot by adding the required electronics and modifying the position of the propeller in order to free the frontal field of view. This required a redesign of the gearbox in order to be able to fit several thin electrical wires in the center of the propeller. When equipped with sensors and electronics, the total weight of the *MC2* reaches 10.3 g (Table 4.2). The airplane is still capable of flying in reasonably small spaces at low velocity (around 1.5 m/s). In this robotic configuration, the average consumption is on the order of 1 W (Table 4.2) and the on-board 65 mAh lithium-polymer battery ensures an energetic autonomy of about 10 minutes.

Table 4.2 Mass and power budgets of the *MC2* microflyer.

Subsystem	Mass [g]	Peak power [mW]
Airframe	1.8	–
Motor, gear, propeller	1.4	800
2 actuators	0.9	450
Lithium-polymer battery	2.0	–
Microcontroller board	1.0	80
Bluetooth radio module	1.0	140
2 cameras with rate gyro	1.8	80
Anemometer	0.4	< 1
Total	10.3	1550

As with the *F2*, no active attitude control is necessary in order for the *MC2* to remain upright during flight. The dihedral of its wing is ensured by a small wire connecting one wing tip with the other and providing a clear tendency towards level attitude. Collision avoidance and altitude control are central issues and the *MC2* possesses enough sensors to cope with both of them, resulting in fully autonomous flight.

	Khepera with kevopic	Indoor airship (Blimp2b)	Indoor airplane (F2)	Indoor microflyer (MC2)
Type	Terrestrial, wheeled	Aerial, buoyant	Aerial, fixed-wing	Aerial, fixed-wing
Degrees of freedom (DOF)	3	4	6	6
Actuators	2 wheels	3 propellers	1 propeller + 2 servos	1 propeller + 2 magnet-in-a-coil
Weight [g]	120	180	30	10
Speed range [m/s]	0 to 0.2	0 to 1	1.2 to 2.5	1 to 2
Test arena size [m]	0.6 × 0.6	5 × 5	16 × 16	6 × 7
Typical power consumption [W]	4	1	1.5	1
Power supply	cable	battery (LiPo)	battery (LiPo)	battery (LiPo)
Energetic autonomy	–	2-3 hours	15-30 minutes	8-10 minutes
Microcontroller board	kevopic	bevopic	pevopic_F2	pevopic_MC2
Vision sensors	1 horizontal	1 horizontal	2 horizontal	1 horizontal, 1 vertical
Rate gyros	1 yaw	1 yaw	1 yaw	1 yaw, 1 pitch
Velocity sensors	wheel encoders	anemometer	–	anemometer
Optic-flow-based strategies (Chap. 6)	Collision avoidance, altitude control (wall following)	–	Course stabilisation, collision avoidance	Course stabilisation, collision avoidance, altitude control
Support evolutionary experiments (Chap. 7)	yes	yes	no	no

Table 4.3 Characteristics of the four robotic platforms.

4.1.4   Comparative Summary of Robotic Platforms

Table 4.3 provides an overview of the four robotic platforms described above. The first part of the table summarises their main characteristics, and the second part contains the on-board electronics and sensors, which are described in the next Section. The last rows show which control strategies are demonstrated in Chapter 6 and which robots are engaged in the evolutionary experiments described in Chapter 7.

Note that this set of four platforms features an increasing dynamic complexity, speed range, and degrees of freedom, allowing the assessment and verification of control strategies and methodologies with an incremental degree of complexity [Zufferey *et al.*, 2003].

4.2   Embedded Electronics

The electronics suite of the robots was conceived to facilitate the transfer of technology and software from one platform to the other. In this Section, the microcontroller boards, the sensors, and the communication systems equipping the four robotic platforms are presented.

4.2.1   Microcontroller Boards

Four similar microcontroller boards were developed (Fig. 4.8), one for each of the four platforms presented above. They can be programmed using the same tools, and the software modules can be easily exchanged among them. A common aspect of these boards is that they are all based on a Microchip™ 8-bit microcontroller. The PIC18F family microcontroller was selected for several reasons. First, PIC18Fs consume only 30-40 mW when running at 20-32 MHz. They support a low voltage (3 V) power supply, which is compatible with single-cell lithium-polymer batteries (3.7 V nominal). They are available in very small packaging (12 $\times$ 12 mm or even 8 $\times$ 8 mm plastic quad flat packages) and therefore have minimal weights ($<$ 0.3 g). Furthermore, PIC18Fs feature a number of integrated hardware peripherals, such as USART (Universal Synchronous Asynchronous Receiver Transmitter), MSSP (Master Synchronous Serial Port, in particular I2C), and ADCs

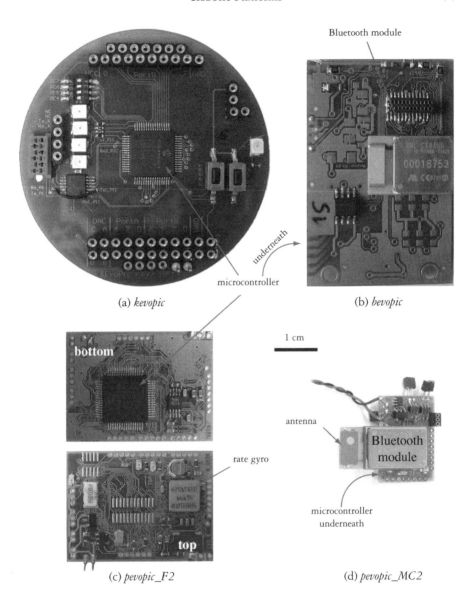

(a) *kevopic*

(b) *bevopic*

(c) *pevopic_F2*

(d) *pevopic_MC2*

Figure 4.8 Microcontroller boards (a) *kevopic* (for the *Khepera*), (b) *bevopic* (for the blimp), (c) *pevopic_F2* and (d) *pevopic_MC2* (for the planes). The microcontrollers are all PIC18F6720 except for the *pevopic_MC2*, which is equipped with a small PIC18F4620. The microcontrollers of the *bevopic* and the *pevopic_MC2* are on the back side of the boards (not visible on the picture). The Bluetooth™ modules with their ceramic antennas are shown only on *bevopic* and *pevopic_MC2*, but are also used on *pevopic_F2*. Also visible on the *pevopic_F2* is an instance of the rate gyro, which is used on all platforms (Sect. 4.2.2).

(Analog to Digital Converters), allowing different types of interfaces with the robot sensors and actuators. The microcontroller can be programmed in assembler as well as in C-language, which enhances the code readability, portability, and modularity.

Naturally, advantages such as low power consumption and small size come at the expense of certain limitations. The PIC18Fs have a reduced instruction set (e.g. 8-bit addition, multiplication, but no division), do not support floating point arithmetic, and feature limited memory (typically 4 kB of RAM, 64 k words of program memory). However, in our approach at controlling indoor flying robots, the limited available processing power is taken as a typical constraint of such platforms. Therefore, the majority of the experiments – at least in their final stage – is performed with embedded software in order to demonstrate the adequacy of the proposed control strategies with truly autonomous, self-contained flying robots.

The microcontroller board for the *Khepera*, the so-called *kevopic*, is not directly connected to some of the robots peripherals (motors, wheel encoders, and proximity sensors), but uses the underlying *Khepera* module as a slave. *Kevopic* has a serial communication link with the underlying *Khepera*, which is only employed for sending motor commands, reading wheel speeds and proximity sensors. The visual and gyroscopic sensors instead are directly connected to *kevopic*, avoiding the transfer of vision stream via the *Khepera* main processor.

The architecture is slightly different for the boards of the flying robots as a result of them being directly interfaced with the sensors and the actuators. In addition to the PIC18F6720 microcontroller, *bevopic* (blimp, evolution, PIC) features three motor drivers and numerous extension connectors, including one for the vision sensor, one for the rate gyro, and one for the remaining sensors and actuators. It is slightly smaller and far lighter than *kevopic* (4.4 g instead of 14 g). It also features a connector for a Bluetooth™ radio module (see Sect. 4.2.3).

The microcontroller board for the *F2* airplane, *pevopic_F2*, is similar to *bevopic*, although much smaller and lighter. *Pevopic_F2* weighs 4 g, the wireless module included, and is half the size of the *bevopic* (Fig. 4.8). This is rendered possible since the servos used on the *F2* do not require bidirectional motor drivers. A simple transistor is sufficient for the main motor and servos for the rudder and the elevator have their own motor drivers. Unlike *bevopic*, *pevopic_F2* has its rate gyro directly on-board in order to avoid the weight of the connection wires and additional electronic board.

The latest version of the microcontroller boards, i.e. *pevopic_MC2*, is less than half the size of *pevopic_F2* and weighs a mere 1 g. It is based on a Microchip, Inc. PIC18LF4620 running at 32 MHz with an internal oscillator, which further reduces the space required for implementing the processor. The board (Fig. 4.9) contains several transistors to directly power the magnet-in-a-coil actuators using PWM signals. It has no onboard rate gyros since these are directly mounted on the back of the cameras (Fig. 4.10).

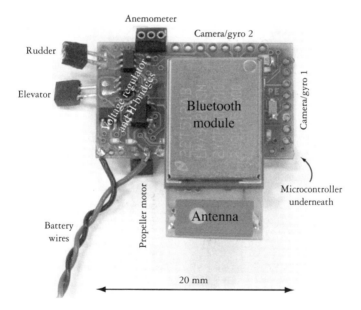

Figure 4.9 A close-up of the *pevopic_MC2* board (1 g) with its piggy-back Bluetooth module (1 g). The connectors to the various peripherals are indicated on the picture.

rate gyro on all four robotic platforms. Modifications were only required with respect to optics and packaging in order to meet the various constraints of the robotic platforms.

Camera and Optics

The selection of a suitable vision system to provide enough information concerning the surrounding environment for autonomous navigation while taking into account the considerable weight constraints of small flying robots is not a trivial task. On the one hand, it is well known that global motion fields spanning a wide field of view (FOV) are easily interpreted [Nelson and Aloimonos, 1988] and indeed most flying insects have an almost omnidirectional vision (see Sect. 3.3.3). On the other hand, artificial

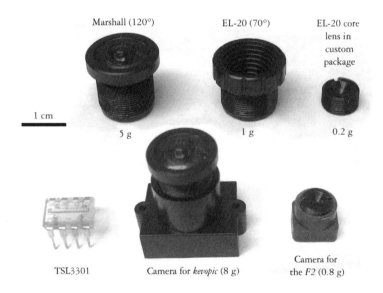

Figure 4.10 (Part a) Cameras for the *Khepera*, *Blimp* and *F2*. The vision chip (bottom-left), optics (top) and camera packaging (bottom center and right). Marshall and EL-20 optics are interchangeable in the camera for *kevopic*. In the effort of miniaturisation, the TSL3301 is machined such to fit into the small custom-developed lens housing labelled "Camera for the *F2*", whose overall size is only $10 \times 10 \times 8$ mm. The 8 pins of the TSL3301 are removed and the chip is directly soldered on the underlying printed circuit board. The EL-20 core plastic lens is extracted from its original packaging and placed into a smaller one (top-right). The weight gain is fivefold (a camera for *kevopic* with an EL-20 weighs 4 g).

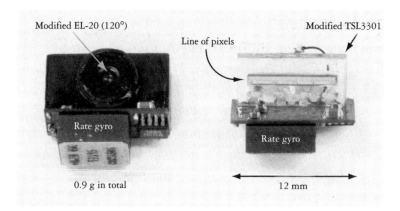

Figure 4.10 (Part b) The camera module for the *MC2*. The latest version for the *MC2* microflyer. Left: The entire module, viewed from the lens side, with a rate gyro soldered underneath the 0.3-mm printed circuit board (PCB). Right: The same module, but without its plastic housing, thus highlighting the underlying TSL3301 that was significantly machined to reduce size and allow vertical soldering on the PCB.

vision systems with wide FOV tend to be heavy due to them needing a special mirror or fish-eye optics with multiple high-quality lenses. Such subsystems are also likely to require much, if not too much, processing power from the on-board microcontroller because of a large number of pixels.

It was therefore decided to use simple, low-resolution, and lightweight 1D cameras (also called *linear* cameras) with lightweight plastic lenses. Such modules can point in different and divergent directions depending on the targeted behaviour. 1D cameras also present the advantage of having few pixels, hence keeping the computational and memory requirements within the limits of a small microcontroller.

The 1D camera that was selected is the Taos Inc. TSL3301 (Fig. 4.10), featuring a linear array of 102 grey-level pixels. However, not all the 102 pixels are usually used either because certain pixels are not exposed by the optics or because only part of the visual field is required for a specific behaviour. Also important is the speed at which images can be acquired. The TSL3301 can be run at a rate as fast as 1 kHz (depending on the exposition time), which is far above what standard camera modules (typically found in mobile phones) are capable of.

Optics and Camera Orientations

In order to focus the light onto the TSL3301 pixel array, two different optics are utilized (Fig. 4.10). The first one, a Marshall-Electronics™ V-4301.9-2.0FT, has a very short focal length of 1.9 mm providing an ultra-large FOV of about $120°$, at the expense of a relatively significant weight of 5 g. The second one, an Applied-Image-group™ EL-20, has a focal length of 3.4 mm and a FOV of approximately $70°$. The advantages of the EL-20 are its relatively low weight (1 g) due to its single plastic lens system and the fact that it can be machined in order for the core lens to be extracted and remounted it in a miniaturised lens-holder weighing only 0.2 g (Fig. 4.10a, top-right). Both optics provide an inter-pixel angle $(1.4\text{-}2.6°)$ comparable to the interommatidial angle in flying insects $(1\text{-}5°$, see Section 3.2.1).

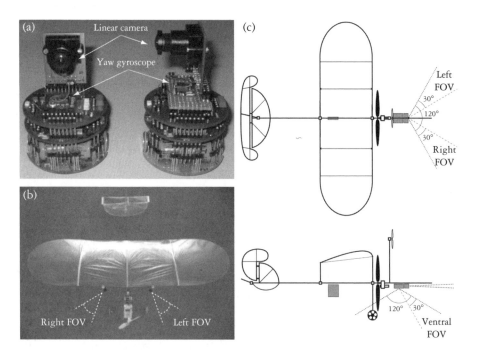

Figure 4.11 The camera position and orientation on the robots (the blimp is not shown here). (a) On the *Khepera*, the camera can be oriented either forward or laterally with a $70°$ or $120°$ FOV depending on the optics (in this picture, the Marshall lens is mounted). (b) A top view of the *F2* showing the orientations of the camera. The FOVs are overlaid in white. (c) Top and side views of the *MC2* with the two FOV of the frontal and ventral cameras. Out of the $2 \times 120°$ FOV, only $3 \times 30°$ are actually used for collision avoidance and altitude control (Chap. 6).

The TSL3301 array of pixels is oriented horizontally on the robots. On the *Khepera*, the camera can be oriented either forward or laterally by adding a small adapter (Fig. 4.11a). On the *Blimp2b*, the camera is mounted at the front end of the gondola and oriented forward (Fig. 4.2). For the experiment described in Chapter 6, the *F2* airplane needs a large FOV, but the weight of a Marshall lens is prohibitive. In fact, the *Khepera* and the *Blimp2b* support both types of lenses, whereas the *F2* is equipped with two miniaturised camera modules, each oriented at $45°$ off the longitudinal axis of the plane (Fig. 4.11b), and featuring the EL-20 as core lens (Fig. 4.10a). The two miniature cameras with custom packaging are indeed tenfold lighter than a single one with a Marshall lens. On the *MC2*, a further optimisation is obtained by removing the cone in front of the EL-20. This modification produces a FOV of $120°$ as a result of the number of pixels exposed to the light increasing from 50 to about 80. Therefore, a single camera pointing forward can replace the two modules present on the *F2* (Fig. 4.11c). A second camera pointing downwards provides a ventral FOV for altitude control.

Rate gyroscope

The Analog-Devices™ ADXRS (Fig. 4.12) is a small and lightweight MEMS (Micro-Electro-Mechanical Systems) rate gyro with very few external components. It consumes only 25 mW but requires a small step-up converter to be powered at 5 V (as opposed to 3.3 V for the rest of the on-board electronics).

Figure 4.12 The ADXRS piezoelectric rate gyro. The ball-grid array (BGA) package is 7×7 mm square, 3 mm thick and weighs 0.4 g.

Very much like the halteres of a fly (see Sect. 3.2.2), such piezoelectric rate gyros rely on the Coriolis effect appearing on vibrating elements to sense the speed of rotation. The ADXRS150 can sense angular velocities up to 150°/s. Taking into account the analog to digital conversion carried out by the microcontroller, the resolution of the system is slightly better than 1°/s over the entire range. Each of our robots are equipped with at least one such rate gyro to measure yaw rotations. The ADXRS on the *Khepera* is visible in Figure 4.11, and the one on the *Blimp2b* is shown in Figure 4.2. The gyro on the *F2* is directly mounted on the *pevopic* board and shown in Figure 4.8c. The *MC2* has two of them, one measuring yawing and the other measuring pitching movements. They were directly mounted on the back of the camera modules (Fig. 4.10b).

Anemometers

The *Blimp2b* and the *MC2* are also equipped with custom-developed anemometers consisting of a free-rotating propeller driving a small magnet in front of a hall-effect sensor (Allegro3212, SIP package) in order to estimate airspeed (the *MC2* version is shown in Figure 4.13). This anemometer is placed in a region that is not blown by the main propeller (see Figure 4.2 for the blimp and Figure 4.7 for the *MC2*). The frequency of the pulsed signal output by the hall-effect sensor is computed by the microcontroller and mapped into an 8-bit variable. This mapping needs to be experimentally tuned in order to fit the typical values obtained in flight.

4.2.3 Communication

In order to monitor the internal state of the robot during the experiments, a communication link that supports bidirectional data transfer in real-time is crucial. In this respect, the *Khepera* is very practical as it can easily be connected to the serial port of a workstation with wires through a rotating contact module (as shown in Figure 4.15a). Of course, this is not possible with the aerial versions of our robots. Thus, to meet the communication requirements, we opted for Bluetooth. Commercially available Bluetooth radio modules are easy to implement and can be directly connected to an RS232 serial port.

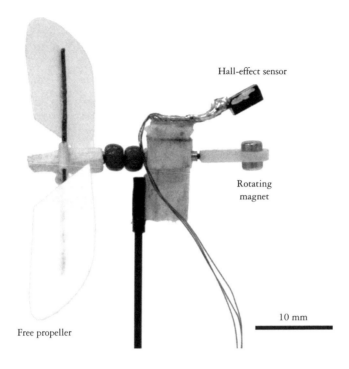

Figure 4.13 The 0.4-gram anemometer equipping the *MC2* is made of a paper propeller linked to a small magnet that rotates in front of a hall-effect sensor.

The selected Bluetooth modules (either the Mitsumi™ WML-C10-AHR, Figure 4.8b, or the National Semiconductor™ LMX9820A, Figure 4.8d) have ceramic antenna and overall weights of only 1 g. They are class 2 modules, which means that the communication range is guaranteed up to 10 m, but in practice distances of up to 25 m in indoor environments present no problems. The power consumption is between 100 to 150 mW during transmission. The more recent LMX9820A emulates a virtual serial communication port without requiring any specific drivers on the host microcontroller. This feature allows for an easy connection to the robot from a Bluetooth-enabled laptop in order to log flight data or reprogram the microcontroller on-board the robot by means of a bootloader.

The advantages of using Bluetooth technology were twofold. Firstly, one can benefit from the continuous efforts toward low power and miniaturisation driven by the market of portable electronic devices. Secondly, Bluetooth modules have several built-in mechanisms to counteract electromagnetic noise, such as frequency hopping and automatic packet retransmission on errors. Therefore, the host microcontroller need not to worry about encoding or error detection and recovery.

To communicate with the robots, a simple packet-based communication protocol is utilized. *Bevopic* and *pevopic* both have connectors supporting either an RS232 cable or a Bluetooth module. When Bluetooth is used, the PIC controls the module via the same serial port. Note that a packet-based protocol is also very convenient for TCP/IP communication, which we employed when working with simulated robots (see Sect. 4.3.2).

4.3   Software Tools

This Section briefly discusses the two main software tools that were used for the experiments described in Chapters 6 and 7. The first is a robot interface and artificial evolution manager used for fast prototyping of control strategies and for evolutionary experiments. The second software is a robot simulator mainly employed for the *Blimp2b*.

4.3.1   Robot Interface

The software *goevo*[3] is a robot interface written in C++ with the wxWidgets[4] framework, to ensure a compatibility with multiple operating systems. *Goevo* implements the simple packet-based protocol (see Sect. 4.2.3) over various kinds of communication channels (RS232, Bluetooth, TCP/IP) in order to receive and send data to/from the robots. It can display sensor data in real-time and log them into text files that can be further analysed with Matlab. It is also very convenient for early stage assessment of sensory-motor loops since control schemes can be easily implemented and assessed on a workstation (which communicates with the real robot at every sensory-motor cycle) before being compiled into the microcontroller firmware for autonomous operation.

---

[3] *goevo* website: http://lis.epfl.ch/resources/evo

[4] wxWidgets website: http://wxwidgets.org/

Goevo can also be used to evolve neural circuits for controlling real or simulated robots. It features built-in neural networks and an evolutionary algorithm (Chap. 7).

4.3.2 Robot Simulator

A robot simulator can be also used to ease the development of control strategies before validating them in real-life conditions. This is particularly useful with evolutionary techniques (Chap. 7) that are known to be time consuming when performed in reality and potentially destructive for the robots.

As a framework for simulating our robots, we employed Webots™ [Michel, 2004], which is a convenient tool for creating and running mobile robot simulations in a 3D environment (relying on OpenGL) with a number of built-in sensors such as cameras, rate gyros, bumpers, range finders, etc. Webots also includes rigid-body dynamics (based on ODE[5]), which provides libraries for kinematic transformations, collision handling, friction and bouncing forces, etc. *Goevo* can communicate with a robot simulated in Webots via a TCP/IP connection, using the same packet-based protocol as employed with the real robots.

The *Khepera* robot, with its wheel encoders and proximity sensors, is readily available in the basic version of Webots. For our experiments, it was augmented with a 1D vision sensor and a rate gyro to emulate the functionality provided by *kevopic*. The test arena was easy to reconstruct using the same textures as employed to print the wallpaper of the real arena.

Webots does not yet support non-rigid-body effects such as aerodynamic or added-mass effects. Thus in order to ensure a realistic simulation of the *Blimp2b*, the dynamic model presented in Zufferey *et al.* [2006] was added as custom dynamics of the simulated robot, while leaving it to Webots to handle friction with walls and bouncing forces when necessary. The custom dynamics implementation takes current velocities and accelerations as input and provides force vectors that are passed to Webots, which computes the resulting new state after a simulation step. Figure 4.14 illustrates

---

[5] Open Dynamics Engine website: http://opende.sourceforge.net

the simulated version of the *Blimp2b*, which features the same set of sensors as its real counterpart (Fig. 4.2). Those sensors are modeled using data recorded from the physical robot. The noise level and noise envelope were reproduced in the simulated sensors to match the real data as closely as possible. In addition to the sensors existing on the physical *Blimp2b*, *virtual sensors*[6] can easily be implemented in simulation. In particular, experiments described in Chapter 7 require the simulated *Blimp2b* to have 8 proximity sensors distributed all around the envelope (Fig. 4.14). This blimp model is now distributed for free with Webots.

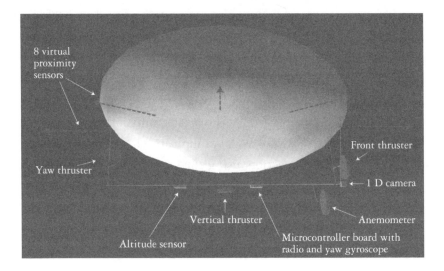

Figure 4.14   A side view of the simulated *Blimp2b*. The darker arrows indicate the direction and range of the virtual proximity sensors.

The simulation rate obtained with all sensors and full physics (built-in and custom) is 40 to 50 times faster than real-time when running on a current PC (e.g. Intel(R) Pentium IV at 2.5 GHz with 512 MB RAM and nVidia(R) GeForce4 graphic accelerator). This rate permitted a significant acceleration of long-lasting experiments such as evolutionary runs.

---

[6] We call "virtual sensors" the sensors that are only implemented in simulation, but do not exist on the real blimp.

4.4   Test Arenas

Since this book is focused on basic, vision-based navigation, the geometry of the test arenas was deliberately kept as simple as possible (Figs 4.15, 4.16 and 4.17). The square textured arenas were inspired by the environ-

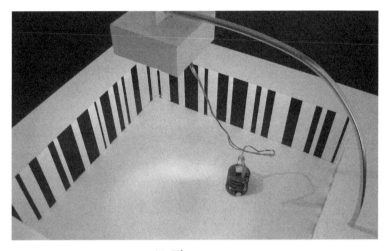

(a) *Khepera* arena

(b) With another pattern

Figure 4.15   Test arenas for the *Khepera*. (a) The *Khepera* arena is 60×60 cm with 30 cm high walls featuring randomly arranged black and white patterns. (b) The same square arena for the *Khepera* with another kind of random pattern on the walls.

ments used in biology for studying visually-guided behaviours in insects (see, Egelhaaf and Borst, 1993a; Srinivasan *et al.*, 1996; Schilstra and van Hateren, 1999; Tammero and Dickinson, 2002a, or Figure 3.8). The simplicity of the shape and textures facilitates the understanding of the principles underlying insect behaviours and the development of the robot control systems.

(a) Real blimp arena

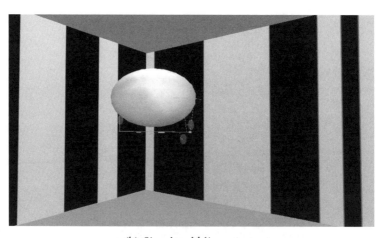

(b) Simulated blimp arena

Figure 4.16 The test arena for the blimp. (a) The blimp arena measures 5 × 5 m and also displays random black and white stripes painted on the walls. (b) The equivalent arena in simulation. The patterns in the simulator were exactly reproduced from the real ones that have been painted on the walls.

Each robot required an arena adapted to its manoeuvrability and velocity. The *Khepera* had a small desktop arena of 60×60 cm, the blimp manoeuvred in a room measuring 5×5 m (3 m high), the *F2* airplane flew in a 16×16 m arena delimited by fabric walls, and the *MC2* manoeuvred in a 7×6 m room equipped with projectors.

In order to provide robots with visual contrast, the walls of these arenas were equipped with random black and white patterns. The random distribution and size of the stripes was intended to ensure that the robots would not rely on trivial geometric solutions to depth perception to navigate. This kind of random distribution was too expensive to be applied in the wide arena for the *F2* (Fig. 4.17a) due to fabric size being standardised and the fact that it would have been too time-consuming to cut and reassemble small pieces of fabric. However, even with this near-homogeneously distributed pattern, the robot did not rely on triangulation to navigate in this room (Chap. 6). The small and more recent test arena for the *MC2* was more flexible (Fig. 4.17b) as it had 8 projectors attached to the ceiling permitting an easy modification of the patterns displayed on the walls.

4.5   Conclusion

This Chapter contained a presentation of the robotic platforms as well as their accompanying tools and test arenas for vision-based navigation experiments described in Chapters 6 and 7. The *Khepera* with *kevopic* is the simplest platform with respect to dynamics and operation since it moves on a flat surface and in a limited space. It can be wired to a computer without the dynamics being affected. For this reason it was used for the preliminary assessment of control strategies. However, this wheeled platform is unable to capture the complex dynamics of flying robots, more specifically their sensitivity to inertia and aerodynamic forces. To test our approach with more dynamic platforms, we developed actual indoor flying robots. In Chapter 6, an autonomous, optic-flow-based aerial steering using the indoor airplanes is described.

(a) *F2* test arena

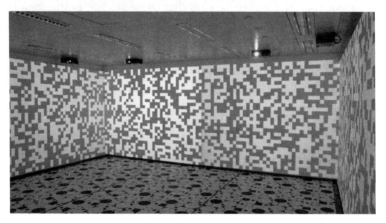

(b) *MC2* test arena

Figure 4.17 Test arenas for the airplanes. (a) The test arena for the *F2* is 16×16 m large and surrounded by soft walls made of fabric. Note that the regularity of the pattern is due to the specific size of the material, but is not imposed by the experiments. (b) The 7×6-m test room for the *MC2* has 8 projectors attached to the ceiling, each projecting on half a wall. This system permitted an easy modification of the textures on the walls. The ground is covered by a randomly textured carpet.

In the search for alternative vision-based navigation strategies, Chapter 7 relies on an evolutionary technique, which presents several difficulties for a robot such as the *F2* or the *MC2*. The strategic decision was thus made to tackle this alternative experimental approach with a more convenient testbed. To that end, we developed the *Blimp2b*, which can fly more easily than an airplane and is able to withstand collisions without breaking down. It also displays simpler dynamics than a plane since critical situations such as stall or aerobatic manoeuvres do not occur with airships. Therefore, an accurate dynamic model of a blimp is simpler to obtain and can be more readily included in a robotic simulator. This is interesting as it allows to significantly speed up the time taken by evolutionary runs.

Obviously, these aerial platforms were not attempts to reproduce the bio-mechanical principles of insect flight. Although, in the future, flapping-wings (see Sect. 2.1) are likely to provide a good alternative for flying in confined environments, they remain mechanically much more complex as well as more delicate to control.

CHAPTER 5

Optic Flow

The real voyage of discovery lies not in seeking new landscapes, but in having new eyes.

M. Proust (1871-1922)

As seen in Chapter 3, the main sensory cue for flight control in insects is visual motion, also called *optic flow* (OF). In this Chapter, the formal description of OF enables us to gain an insight of global motion fields generated by particular movements of the observer and the 3D structure of the scene. It is then possible to analyse them and develop the control strategies presented in the following Chapter. In practice, lightweight robots cannot afford high-resolution, omnidirectional cameras and computationally intensive algorithms. OF must thus be estimated with limited resources in terms of processing power and vision sensors. In the second part of this Chapter, an algorithm for OF detection that meets the constraints imposed by the 8-bit microcontroller equipping our robots is described. Combining this algorithm with a 1D camera results in what we call an *optic-flow detector* (OFD). Such an OFD is capable of measuring OF in real-time along one direction in a selectable part of the field of view. Several of these OFDs, spanning one or more cameras are implemented on the robots to serve as image preprocessors for navigation control.

5.1   What is Optic Flow?

Optic flow is the perceived visual motion of objects as the observer moves relative to them. It is generally very useful for navigation because it contains information regarding self-motion and the 3D structure of the environment. The fact that visual perception of changes represents a rich source of information about the world has been widely spread by Gibson [1950]. In this book, a stationary environment is assumed in order for the optic flow to be generated solely by the self-motion of the observer.

5.1.1   Motion Field and Optic Flow

In general, a difference has to be made between *motion field* (sometimes also called *velocity field*) and *optic flow* (or *optical flow*). The motion field is the 2D projection onto a retina of the relative 3D motion of scene points. It is thus a purely geometrical concept, and has nothing to do with image intensities. On the other hand, the optic flow is defined as the apparent motion of the image intensities (or brightness patterns). Ideally, the optic flow corresponds to the motion field, but this may not always be the case [Horn, 1986]. The main reasons for discrepancies between optic flow and motion field are the possible absence of brightness gradients or the aperture problem[1].

In this project, however, we deliberately confound these two notions. In fact, there is no need, from a behavioural perspective, to rely on the ideal motion field. It is sufficient to know that the perceived optic flow tends to follow the main characteristics of the motion field (such as an increase when approaching objects). This was very likely to be the case in our test environments where significant visual contrast was available (Sect. 4.4). Moreover, spatial and temporal averaging can be used (as in biological systems) to smooth out perturbations arising in small parts of the visual field where no image patterns would be present for a short period of time.

_____

[1] If the motion of an oriented element is detected by a unit that has a small FOV compared to the size of the moving element, the only information that can be extracted is the component of the motion perpendicular to the local orientation of the element [Marr, 1982, p.165; Mallot, 2000, p.182].

In addition, there is always a difference between the actual optic flow arising on the retina and the one a specific algorithm measures. However, our simple robots are not intended to retrieve metric information about the surrounding world, but rather use qualitative properties of optic flow to navigate. Relying on rough optic-flow values for achieving efficient behaviours rather than trying to estimate accurate distances is indeed what flying insects are believed to do [Srinivasan *et al.*, 2000]. There is also good evidence that flies do not solve the aperture problem, at least not at the level of the tangential cells [Borst *et al.*, 1993].

In Chapter 6, the formal description of the motion field is used in order to build ideal optic-flow fields arising in particular flight situations to draw conclusions about the typical flow patterns that can be used for implementing basic control strategies like collision avoidance and altitude control. Unlike the eyes of flying insects, the cameras of our robots have limited FOV (see Sect. 4.2.2), and this qualitative study thus provides a basis for deciding in which directions the cameras, and thereby also the OFDs, should be oriented.

5.1.2 Formal Description and Properties

Here, the formal definition of optic flow (as if it were identical to the motion field) is discussed and interesting properties are highlighted.

A vision sensor moving within a 3D environment ideally produces a time-varying image which can be characterised by a 2D vector field of local velocities. This motion field describes the 2D projection of the 3D motion of scene points relative to the vision sensor. In general, the motion field depends on the motion of the vision sensor, the structure of the environment (distances to objects), and the motion of objects in the environment, which are assumed to be null in our case (stationary environment).

For the sake of simplicity, we consider a spherical visual sensor of unit radius[2] (Fig. 5.1). The image is formed by spherical projection of the environment onto this sphere. Apart from resembling the case of a fly's eye,

---

[2] A unit radius allows the normalisation the OF vectors on its surface and the expression of their amplitude directly in [rad/s].

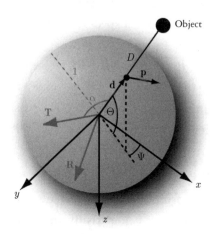

Figure 5.1 The spherical model of a visual sensor. A viewing direction indicated by the unit vector **d**, which is a function of azimuth Ψ and elevation Θ (spherical coordinates). The distance to an object in the direction $\mathbf{d}(\Psi, \Theta)$ is denoted $D(\Psi, \Theta)$. The optic-flow vectors $\mathbf{p}(\Psi, \Theta)$ are always tangential to the sphere surface. The vectors **T** and **R** represent the translation and rotation of the visual sensor with respect to its environment. As will be seen in the next Section, the angle α between the direction of translation and a specific viewing direction is sometimes called *eccentricity*.

the use of a spherical projection makes all points in the image geometrically equivalent, thus simplifying the mathematical analysis[3]. The photoreceptors of the vision sensor are thus assumed to be arranged on this unit sphere, each photoreceptor defining a viewing direction indicated by the unit vector $\mathbf{d}(\Psi, \Theta)$, which is a function of both azimuth Ψ and elevation Θ in spherical coordinates. The 3D motion of this vision sensor can be fully described by a translation vector **T** and a rotation vector **R** (describing the axis of rotation and its amplitude)[4]. When the vision sensor moves in its environment, the motion field $\mathbf{p}(\Psi, \Theta)$ is given by Koenderink and van Doorn [1987]:

---

[3] Ordinary cameras do not use spherical projection. However, if the FOV is not too wide, this approximation is reasonably close [Nelson and Aloimonos, 1989]. A direct model for planar retinas can be found in Fermüller and Aloimonos [1997].

[4] In the case of an aircraft, **T** is a combination of thrust, slip, and lift, and **R** a combination of roll, pitch, and yaw.

$$\mathbf{p}(\Psi,\Theta) = \left[-\frac{\mathbf{T} - \big(\mathbf{T}\cdot\mathbf{d}(\Psi,\Theta)\big)\mathbf{d}(\Psi,\Theta)}{D(\Psi,\Theta)} \right] \qquad (5.1)$$
$$+ \left[-\mathbf{R}\times\mathbf{d}(\Psi,\Theta) \right],$$

where $D(\Psi,\Theta)$ is the distance between the sensor and the object seen in direction $\mathbf{d}(\Psi,\Theta)$. Although $\mathbf{p}(\Psi,\Theta)$ is a 3D vector field, it is by construction tangential to the spherical sensor surface. Optic-flow fields are thus generally represented by the unfolding of the spherical surface into a Mercator map (Fig. 5.2). Positions in the 2D space of such maps are also defined by the azimuth Ψ and elevation Θ angles.

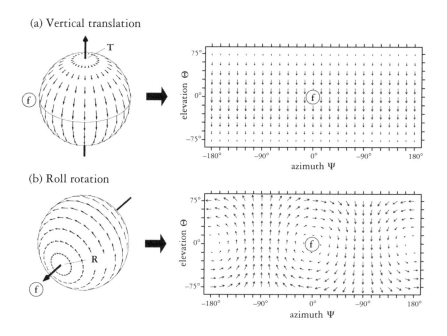

Figure 5.2 Optic-flow fields due to (a) an vertical translation and (b) a rotation around the roll axis. The projection of the 3D relative motion on spherical visual sensors (left) and the development of the sphere surface into Mercator maps (right). The encircled "f" indicates the forward direction. Reprinted with permission from Dr Holger Krapp.

Given a particular self-motion **T** and **R**, along with a specific repartition of distances $D(\Psi, \Theta)$ to surrounding objects, equation (5.1) allows the
reconstruction of the resulting theoretical optic-flow field. Beyond that, it
formally supports a fact that was already suggested in Chapter 3, i.e. that
the optic flow is a linear combination of the translational and rotational
components[5] induced by the respective motion along **T** and around **R**.
The first component, hereafter denoted *TransOF*, is due to translation and
depends on the distance distribution, while the second component, *RotOF*,
is produced by rotation and is totally independent of distances (Fig. 5.3).

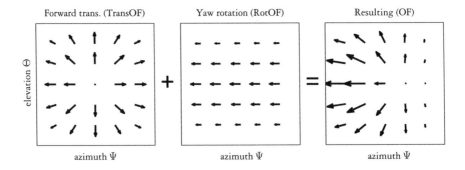

Figure 5.3 OF fields showing the effect of the superposition of TransOF and
RotOF. The hypothetical camera is oriented toward a fronto-parallel plane. The
first OF field is due to forward translation whereas the second one is a result of yaw
rotation.

From equation (5.1) it can be seen that the TransOF amplitude is inversely proportional to distances $D(\Psi, \Theta)$. Therefore, if the translation is
known and the rotation is null, it is in principle possible to estimate the distances to surrounding objects. In free-manoeuvring agents, however, the rotational and translational optic-flow components are linearly superimposed
and may result in rather complex optic-flow fields. It is quite common that
RotOF overwhelms TransOF, thus rendering an estimation of the distance

---

[5] The local flow vectors in translational OF fields are oriented along meridians connecting the focus of expansion (FOE, i.e. the direction point in which the translation is
pointing) with the focus of contraction (FOC, which is the opposite pole of the flow
field). A general feature of the RotOF structure is that all local vectors are aligned
along parallel circles centered around the axis of rotation.

quite difficult. This is probably the reason why flies tend to fly straight and actively compensate for unwanted rotations (see Sect. 3.4.3). Another means of compensating for the spurious RotOF signals consists in deducing it from the global flow field by measuring the current rotation with another sensory modality such as a rate gyro. Such a process is often called *derotation*. Although this solution has not been shown to exist in insects, it an efficient way of avoiding active gaze stabilization mechanisms in robots.

5.1.3 Motion Parallax

A particular case of the general equation of optic flow (5.1) is often used in biology [Sobel, 1990; Horridge, 1977] and robotics [Franceschini *et al.*, 1992; Sobey, 1994; Weber *et al.*, 1997; Lichtensteiger and Eggenberger, 1999] to explain depth perception from optic flow. The so-called *motion parallax* refers to a planar situation where only pure translational motion is (Fig. 5.4). In this case, it is trivial[6] to express the optic-flow amplitude p (also referred to as the *apparent angular velocity*) provoked by an object at distance D, seen at an angle α with respect to the motion direction $\mathbf{T}$:

$$p(\alpha) = \frac{\|\mathbf{T}\|}{D(\alpha)} \sin \alpha, \qquad \text{where } p = \|\mathbf{p}\|. \qquad (5.2)$$

Note that if $\mathbf{T}$ is aligned with the center of the vision system, the angle α is often called *eccentricity*. The formula was first derived by Whiteside and Samuel, 1970 in a brief paper concerning the blur zone that surrounds an aircraft flying at low altitude and high speed. If the translational velocity and the optic-flow amplitude are known, the distance from the object can thus be retrieved as follows:

$$D(\alpha) = \frac{\|\mathbf{T}\|}{p(\alpha)} \sin \alpha. \qquad (5.3)$$

---

[6] To derive the motion parallax equation (5.2) from the general optic-flow equation (5.1), the rotational component must first be cancelled since no rotation occurs, subsequently, the translation vector $\mathbf{T}$ should be expressed in the orthogonal basis formed by $\mathbf{d}$ (the viewing direction) and $\frac{\mathbf{p}}{\|\mathbf{p}\|}$ (the normalised optic-flow vector).

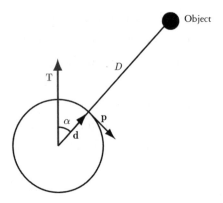

Figure 5.4 The motion parallax. The circle represents the retina of a moving observer. The symbols are defined in Figure 5.1.

The motion parallax equation (5.2) is interesting in the sense that it gives a sense of how the optic flow varies on the retina depending on the motion direction and the distance to objects at various eccentricities.

5.2   Optic Flow Detection

Whereas the previous Section provides an overview on ideal optic-flow fields, the objective here is to find an optic-flow algorithm that would eventually lead to an implementation on the available hardware.

5.2.1   Issues with Elementary Motion Detectors

Following a bio-inspired approach, the most natural method for detecting optic flow would be to use correlation-type EMDs[7]. However, beyond the fact that EMD models are still subject to debate in biology and their spatial integration is not yet totally understood (see Sect. 3.3.2), the need for true image velocity estimates and insensitivity to contrast and spatial frequency of visual surroundings led us to turn away from this model.

---

[7] In fact, it is possible to implement several real-time correlation-type EMDs (e.g. Iida and Lambrinos, 2000) running on a PIC microcontroller. However, the filter parameter tuning is tedious and, as expected, the EMD response is non-linear with respect to image velocity and strongly depends on image contrast.

It is often proposed (e.g. Harrison and Koch, 1999; Neumann and Bülthoff, 2002; Reiser and Dickinson, 2003) to linearly sum EMD signals over large receptive fields in order to smooth out the effect of non-linearities and other imprecisions. However, a linear spatial summation can produce good results only if a significant amount of detectable contrast is present in the image. Otherwise the spatial summation is highly dependent on the number of intensity changes (edges) capable of triggering an EMD signal. In our vertically striped test arenas (Sect. 4.4), the spatial summation of EMDs would be highly dependent on the number of viewed edges, which is itself strongly correlated with the distance from the walls. Even with a random distribution of stripes or blobs, there is indeed more chance of seeing several edges from far away than up close. As a result, even if a triggered EMD tends to display an increasing output with decreasing distances, the number of active EMDs in the field of view simultaneously decreases. In such cases, the linear summation of EMDs hampers the possibility of accurately estimating distances.

Although a linear spatial pooling scheme is suggested by the matched-filter model of the tangential cells (see Fig. 3.10) and has been used in several robotic projects (e.g. Neumann and Bülthoff, 2002; Franz and Chahl, 2002; Reiser and Dickinson, 2003), linear spatial integration of EMDs is not an exact representation of what happens in the flies tangential cells (see Sect. 3.3.3). Conversely, important non-linearities have been highlighted by several biologists [Hausen, 1982; Franceschini *et al.*, 1989; Haag *et al.*, 1992; Single *et al.*, 1997], but are not yet totally understood.

5.2.2 Gradient-based Methods

An alternative class of optic-flow computation has been developed within the computer-vision community (see Barron *et al.*, 1994; Verri *et al.*, 1992 for reviews). These methods can produce results that are largely independent of contrast or image structure.

The standard approaches, the co-called *gradient-based methods* [Horn, 1986; Fennema and Thompson, 1979; Horn and Schunck, 1981; Nagel, 1982], assume that the brightness (or intensity) $I(n, m, t)$ of the image of a point in the scene does not change as the observer moves relative to it, i.e.:

$$\frac{dI(n, m, t)}{dt} = 0. \qquad (5.4)$$

Here, n and m are respectively the vertical and horizontal spatial coordinates in the image plane, and t is the time. Equation (5.4) can be rewritten as a Taylor series. Simple algorithms throw away the second order derivatives. In the limit as the time step tends to zero, we obtain the so-called *optic flow constraint equation*:

$$\frac{\partial I}{\partial n}\frac{dn}{dt} + \frac{\partial I}{\partial m}\frac{dm}{dt} + \frac{\partial I}{\partial t} = 0, \qquad \text{with } \mathbf{p} = \left(\frac{dn}{dt}, \frac{dm}{dt} \right). \qquad (5.5)$$

Since this optic flow constraint is a single linear equation in two unknowns, the calculation of the 2D optic-flow vector $\mathbf{p}$ is underdetermined. To solve this problem, one can introduce other constraints like, e.g. the *smoothness constraint* [Horn and Schunck, 1981; Nagel, 1982] or the assumption of *local constancy*[8]. Despite their differences, many of the gradient-based techniques can be viewed in terms of three stages of processing [Barron et al., 1994]: (i) prefiltering or smoothing, (ii) computation of spatiotemporal derivatives, and (iii) integration of these measurements to produce a two-dimensional flow field, which often involves assumptions concerning the smoothness. Some of these stages often rely on iterative processes. As a result, the gradient-based schemes tend to be computationally intensive and very few of them are able to support real-time performance [Camus, 1995].

Srinivasan [1994] has proposed an *image interpolation algorithm*[9] (I2A) in which the parameters of global motion in a given region of the image

---

[8] The assumption that the flow does not change significantly in small neighbourhoods (local constancy of motion).

[9] This technique is quite close to the image registration idea proposed by Lucas and Kanade [1981]. I2A has been further developed by Bab-Hadiashar et al. [1996], who quotes a similar methodology by Cafforio and Rocca [1976]. A series of applications using this technique (in particular for self-motion computation) exists [Chahl and Srinivasan, 1996; Nagle and Srinivasan, 1996; Franz and Chahl, 2002; Chahl et al., 2004]. The I2A abbreviation is due to Chahl et al. [2004].

can be estimated by a single-stage, non-iterative process. This process interpolates the position of a newly acquired image in relation to a set of older reference images. This technique is loosely related to a gradient-based method, but is superior to it in terms of its robustness to noise. The reason for this is that, unlike the gradient scheme that solves the optic flow constraint equation (5.5), the I2A incorporates an error-minimising strategy.

As opposed to spatially integrating local measurements, the I2A estimates the global motion of a whole image region covering a wider FOV (Fig. 5.5). Unlike spatially integrated EMDs, the I2A output thus displays no dependency on image contrast, nor on spatial frequency, as long as some image gradient is present somewhere in the considered image region.

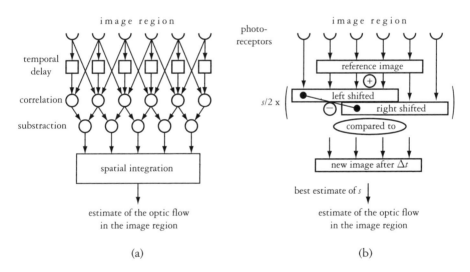

(a)   (b)

Figure 5.5 An EMD vs I2A comparison (unidimensional case). (a) The spatial integration of several elementary motion detectors (EMDs) over an image region. See Figure 3.7 for details concerning the internal functioning of an EMD. (b) The simplified image interpolation algorithm (I2A) applied to an image region. Note that the addition and subtraction operators are pixel-wise. The symbol s denotes the image shift along the 1D array of photoreceptors. See Section 5.2.3 for details on the I2A principle.

5.2.3   Simplified Image Interpolation Algorithm

To meet the constraints of our hardware, the I2A needs to be adapted to 1D images and limited to pure shifts (image expansion or other deformations are not taken into account in this simplified algorithm). The implemented algorithm works as follows (see also Fig. 5.5b). Let $I(n)$ denote the grey level of the nth pixel in the 1D image array. The algorithm computes the amplitude of the translation s between an image region (hereafter simply referred to as the "image") $I(n, t)$ captured at time t, called *reference image*, and a later image $I(n, t + \Delta t)$ captured after a small period of time Δt. It assumes that, for small displacements of the image, $I(n, t + \Delta t)$ can be approximated by $\hat{I}(n, t + \Delta t)$, which is a weighted linear combination of the reference image and of two shifted versions $I(n \pm k, t)$ of that same image:

$$\hat{I}(n, t + \Delta t) = I(n, t) + s\frac{I(n - k, t) - I(n + k, t)}{2k}, \qquad (5.6)$$

where k is a small reference shift in pixels. The image displacement s is then computed by minimising the mean square error E between the estimated image $\hat{I}(n, t + \Delta T)$ and the new image $I(n, t + \Delta t)$ with respect to s:

$$E = \sum_n \left(I(n, t + \Delta t) - \hat{I}(n, t + \Delta t) \right)^2, \qquad (5.7)$$

$$\frac{dE}{ds} = 0$$

$$\Leftrightarrow s = 2k\frac{\sum_n \left(I(n, t + \Delta t) - I(n, t) \right) \left(I(n - k, t) - I(n + k, t) \right)}{\sum_n \left(I(n - k, t) - I(n + k, t) \right)^2}. $$

$$(5.8)$$

In our case, the shift amplitude k is set to 1 pixel and the delay Δt is such to ensure that the actual shift does not exceed ± 1 pixel. $I(n \pm 1, t)$ are thus artificially generated by translating the reference image by one pixel to the left and to the right, respectively.

Note that in this restricted version of the I2A, the image velocity is assumed to be constant over the considered region. Therefore, in order to measure non-constant optic-flow fields, I2A must be applied to several sub-regions of the image where the optic flow can be considered constant. In practice, the implemented algorithm is robust to small deviations from this assumption, but naturally becomes totally confused if opposite optic-flow vectors occur in the same image region.

In the following, the software (I2A) and hardware (a subpart of the 1D camera pixels) are referred to as an optic-flow detector (OFD). Such an OFD differs from an EMD in several respects. It has generally a wider FOV that can be adapted (by changing the optics and/or the number of pixels) to the expected structure of the flow field. In some sense, it participates in the process of spatial integration by relying on more than two neighboring photoreceptors. However, it should always do so in a region of reasonably constant OF. In principle, it has no dependency on contrast or on spatial frequency of the image and its output displays a good linearity with respect to image velocity as long as the image shift remains within the limit of one pixel, or k pixels, in the general case of equation (5.8).

5.2.4 Algorithm Assessment

In order to assess this algorithm with respect to situations that could be encountered in real-world conditions, a series of measurements using artificially generated 1D images were performed in which the I2A output signal s was compared to the actual shift of the images. A set of high-resolution, sinusoidal, 1D gratings were generated and subsampled to produce 50-pixel-wide images with various shifts from -1 to $+1$ pixel with 0.1 steps. The first column of Figure 5.6 shows sample images from the series of artificially generated images without perturbation (case A) and with maximal perturbation (case B). The first line of each graph corresponds to the I2A reference image whereas the following ones represent the shifted versions of the reference image. The second column of Figure 5.6 displays the OF estimation produced by I2A versus the actual image shift (black lines) and the error E (equation 5.7) between the best estimate images and the actual ones (gray lines). If I2A is perfect at estimating the true shift, the black line should correspond to the diagonal. The third column of Figure 5.6 highlights the quality of the OF estimate (mean square error) with

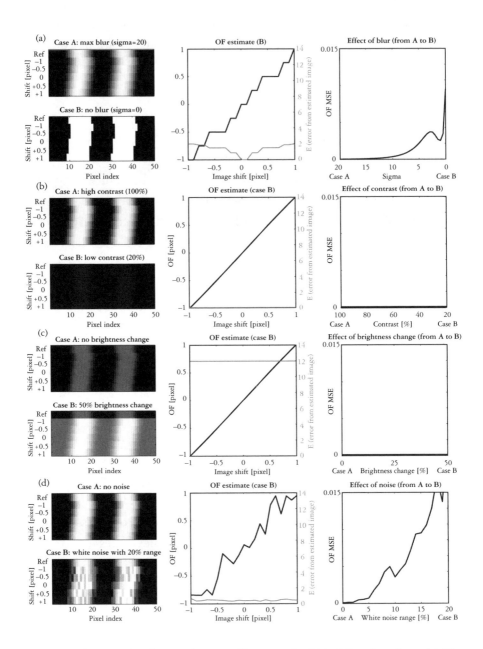

Figure 5.6 A study of perturbation effects on the I2A OF estimation. (a) The effect of Gaussian blur (sigma is the filter parameter). (b) The effect of contrast. (c) The effect of a change in brightness between reference image and the new image. (d) The effect of noise. See text for details.

respect to the degree of perturbation (from case A to case B). In this column, a large OF mean square error (MSE) indicates a poor OF estimation.

A first issue concerns the sharpness of the image. In OF estimation, it is customary to preprocess images with a spatial low-pass filter in order to cancel out high-frequency content and reduce the risk of aliasing effects. This holds true also for I2A and Figure 5.6(a) shows the poor quality of an OF estimation with binary images (i.e. only totally black or white pixels). This result was expected since the spatial interpolation is based on a first-order numerical differentiation, which fails to provide a good estimate of the slope in presence of discontinuities (infinite slopes). It is therefore important to low-pass filter images such that the edges are spread over several adjacent pixels. A trade-off has to be found, however, between binary images and totally blurred ones where no gradient can be detected. A clever way to obtain low-pass filtered images at no computational cost is to slightly defocus the optics.

A low contrast[10] does not alter the I2A estimates (Fig. 5.6b). As long as the contrast is not null, OF computation can be reliably performed. This means that for a given image, there is almost no dependency on brightness settings of the camera, as long as the image gradient is not null. As a result, one can easily find a good exposure time setting and automatic brightness adjustment mechanisms could be avoided in most cases. Note that this analysis does not take noise into account and it is likely that noisy images will benefit from higher contrast in order to disambiguate real motion from spurious motion due to noise.

Another issue with simple cameras in artificially lit environments consists in the flickering or light due to AC power sources, which could generate considerable change in brightness between two successive image acquisitions of the I2A. Figure 5.6(c) shows what happens when the reference image is dark and the new image is up to $50\,\%$ brighter. Here too, the algorithm performs very well, although, as could be expected, the error E is very large as compared to the other cases. This means that even if the best estimated image $\hat{I}(n, t + \Delta t)$ is far from the actual new image because of

---

[10] Contrast is taken in the sense of the absolute difference between the maximum and minimum intensities of an image.

the global difference in brightness, it is still the one that best matches the actual shift between $I(n, t)$ and $I(n, t + \Delta t)$.

Another potential perturbation is the noise that can occur independently on each pixel (due to electrical noise within the vision chip or local optical perturbations). This has been implemented by the superposition of a white noise up to 20 % in intensity to every pixel of the second image (Fig. 5.6d). The right-most graph shows that such a disturbance has a minor effect up to 5 %, while the center graph demonstrates the still qualitatively consistent although noisy OF estimate even with 20 %. Although I2A is robust with respect to a certain amount of noise, significant random perturbations, such as those arising when part of the camera is suddenly saturated due to a lamp or a light reflection entering the field of view, may significantly affect its output. A temporal low-pass filter is thus implemented, which helps to cancel out such spurious data.

The results can be summarised as follows. This technique for estimating OF has no dependency on contrast as long as some image gradient can be detected. The camera should be slightly defocused to implement a spatial low-pass filter. Finally, flickering due to artificial lighting does not present an issue.

5.2.5   Implementation Issues

In order to build an OFD, equation (5.8) must be implemented in the embedded microcontroller, which grabs two successive images corresponding to $I(n, t)$ and $I(n, t + \Delta t)$ with a delay of a few milliseconds (typically 5-15 ms) at the beginning of every sensory-motor cycle. Pixel intensities are encoded on 8 bits, whereas other variables containing the temporal and spatial differences are stored in 32-bit integers. For every pixel, equation (5.8) requires only two additions, two subtractions and one multiplication. These operations are included in the instruction set of the PIC18F microcontroller and can thus be executed very efficiently even with 32-bit integers. The only division of the equation occurs once per image region, at the end of the accumulation of the numerator and denominator. Since the programming is carried out in C, this 32-bit division relies on a compiler built-in routine, which is executed in a reasonable amount of time since the entire computation for a region of 30 pixels is performed within 0.9 ms. As a comparison, a typical sensory-motor cycle lasts between 50 and 100 ms.

In order to assess the OFD output in real-world conditions, the I2A algorithm was first implemented on the PIC of *kevopic* equipped with the frontal camera (see Sect. 4.2.2) and mounted on a *Khepera*. The *Khepera* was then placed in the 60×60 cm arena (Fig 4.15a) and programmed to rotate on the spot at various speeds. In this experiment, the output of the OFD can be directly compared to the output of the rate gyro. Figure 5.7(a) presents the results obtained from an OFD with an image region of 48 pixels roughly spanning a $120°$ FOV. Graph (a) illustrates the perfect linearity of the OF estimates with respect to the robot rotation speed. This linearity is in strong contrast with what could be expected from EMDs (see Figure 3.8 for comparison). Even more striking is the similarity of the standard deviations between the rate gyro and OFD. This indicates that most of the noise, which is indeed very small, can be explained by mechanical vibrations

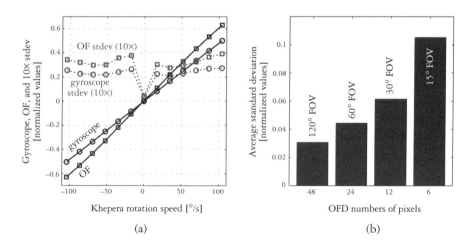

(a)   (b)

Figure 5.7 An experiment with a purely rotating *Khepera* in order to compare I2A output with gyroscopic data. The sensor signals are normalised with respect to the entire range of a signed 8-bit integer (± 127). (a) Rate gyro data (solid line with circles), the related standard deviation of 1000 measurements for each rotation speed (dashed line with circles), and OF values estimated using 48 pixels (solid line with squares), the related standard deviation (dashed line with squares). A value of 0.5 for the rate gyro corresponds to $100°/s$. The optic flow scale is arbitrary. (b) The average standard deviation of OF as a function of the FOV and corresponding pixel number.

of the *Khepera* (which is also why the standard deviation is close to null at $0°/s$), and that the OFD is almost as good as the rate gyro at estimating rotational velocities. This result support our earlier suggestion concerning the derotation of optic flow by simply subtracting a scaled version of the rate gyro output from the global OF. Note that rather than scaling the OFD output, one can simply adjust the delay Δt between the acquisition of the two successive images of I2A so as to match the gyroscopic values in pure rotation.

Field of View and Number of Pixels

To assess the effects of the FOV on the accuracy of an OFD output, the same experiment was repeated while varying the number of pixels. For a given lens, the number of pixels is indeed directly proportional to the FOV. The $120°$ lens (Marshall) used in this experiment induced a low angular resolution. The results shown here thus represent the worst case, since the higher the resolution, the better was the accuracy of the estimation. Figure 5.7(b) shows the average standard deviation of the OF measurements. The accuracy decreases but remains reasonable up to 12 pixels and $30°$ FOV. With only 6 pixels and $15°$, the accuracy is a third of the value with 48 pixels. This trend can be explained by the discretisation errors having a lower impact with large amounts of pixels. Another factor is that a wider FOV provides richer images with more patterns allowing for a better match of the shifted images. At the limit, a too small FOV would sometimes have no contrast at all in the sampled image. When using such OFDs, a trade-off needs to be found between a large enough FOV in order to ensure good accuracy and a small enough FOV in order to better meet the assumption of local constancy of motion; this when the robot is not undergoing only pure rotations.

To ensure that this approach continues to provide good results with another optics and in another environment, we implemented two OFDs on the *F2* airplane, one per camera (see Figure 4.11(b) for the camera orientations). This time, a FOV of $40°$ per OFD was chosen, which corresponds to 28 pixels with the EL-20 lens. The delay Δt was adjusted to match the rate gyro output in pure rotation. The calibration provided an optimal Δt of 6.4 ms. The airplane was then handled by hand and rotated about its yaw axis in its test arena (Fig. 4.17a). Figure 5.8 shows the data recorded during this operation and further demonstrates the good match between rotations estimated by the two OFDs and by the rate gyro.

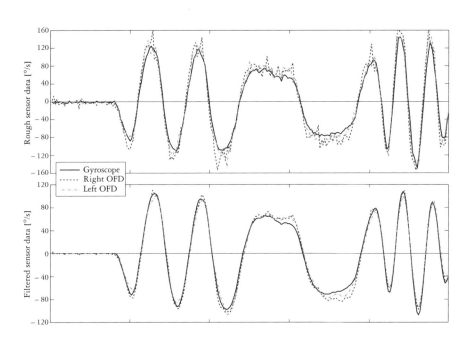

Figure 5.8 A comparison of the rate gyro signal with the estimates of OFDs in pure rotation. The data was recorded every 80 ms while the *F2* was held by hand in the test arena and randomly rotated around its yaw axis. The top graph displays the raw measurements, whereas the bottom graph shows their low-pass filtered version. 100°/s is approximately the maximum rotation speed of the plane in flight.

Optic-flow Derotation

Since RotOF components do not contain any information about surrounding distances, for all kinds of tasks related to distance estimation a pure translational OF field is desirable [Srinivasan *et al.*, 1996]. This holds true for the robots just as it does for flies, which are known to compensate for rotations with there head (see Sect. 3.4.2). Since our robots cannot afford additional actuators to pan and tilt there visual system, a purely computational way of derotating optic flow is used.

It is in principle possible to deduce RotOF from the global flow field by simple vector subtraction, since the global OF is a linear combination of translational and rotational components (Sect. 5.1.2). To do so, it is necessary to know the rotation rate, which can be measured be rate gyros. In our case the situation is quite trivial because the OFDs are unidimensional and the rate gyros have always been mounted with there axes oriented perpendicular to the pixel array and the viewing direction of the corresponding camera (see Sect. 4.2.2). This arrangement reduces the correction operation to a scalar subtraction. Of course a simple subtraction can be used only if the optic-flow detection is linearly dependent on the rotation speed, which is indeed the case of our OFDs (as opposed to EMDs). Figure 5.8 further supports this method of OF derotation by demonstrating the good match between OFD signals and rate gyro output in pure rotation.

5.3   Conclusion

The first Section of this Chapter provided mathematical tools (equations 5.1 and 5.2) used to derive the amplitude and direction of optic flow given the self-motion of the agent and the geometry of the environment. These tools will be of great help in Chapter 6 both to decide how to orient the OFDs and to devise control strategies using their outputs. Another important outcome of the formal description of optic flow is its linear separability into a translational component (TransOF) and a rotational component (RotOF). Only TransOF provides useful information concerning the distance to objects.

The second Section presented the implementation of an optic-flow detector (OFD) that fits the hardware constraints of the flying platforms while featuring a linear response with respect to image velocity. Several of these can be implemented on a small robot, each considering different parts of the FOV (note that they could even have overlapping receptive fields) where the optic flow is assumed to be coherent (approximately the same amplitude and direction).

CHAPTER 6

Optic-flow-based Control Strategies

When we try to build autonomous robots, they are almost literally puppets acting to illustrate our current myths about cognition.

I. Harvey, 2000

This Chapter describes the development and assessment of control strategies for autonomous flight. It is now assumed that the problem concerning local optic-flow detection is solved (Chap. 5) and we thus move on to the question of how these signals can be combined in order to steer a flying robot. This Chapter focuses on spatial combinations of optic-flow signals and integration of gyroscopic information so as to obtain safe behaviours: to remain airborne while avoiding collisions. Since the problem is not that trivial, at least not from an experimental point of view, we proceed step by step. First, the problem of collision avoidance is considered as a 2D steering problem assuming the use of only the rudder of the aircraft while altitude is controlled manually through a joystick connected to the elevator of the airplane. Then, the problem of controlling the altitude is tackled by using ventral optic-flow signals. By merging lateral steering and altitude control, we hope to obtain a fully autonomous system. It turns out, however, that the merging of these two control strategies is far from straightforward. Therefore, the last Section proposes a slightly different approach in which we consider both the walls and the ground as obstacles that must be avoided without distinction. This approach ultimately leads to a fully autonomous system capable of full 3D collision avoidance.

6.1   Steering Control

Throughout this Section, it is assumed that the airplane can fly at constant altitude and we are not preoccupied with how this can be achieved. The problem is therefore limited to 2D and the focus is put on how collisions with walls can be avoided. To understand how optic-flow signals can be combined to control the rudder and steer the airplane, we consider concrete cases of optic-flow fields arising in typical phases of flight. This analysis also allows us to answer the question of where to look, and thus define the orientation of the OFDs.

6.1.1   Analysis of Frontal Optic Flow Patterns

By using equation (5.1) one can easily reconstruct the optic-flow (OF) patterns that arise when an airplane approaches a wall. Since the rotational optic flow (RotOF) does not contain any information about distances, this Section focuses exclusively on translational motion. In practice, this is not a limitation since we showed in the previous Chapter that OF can be derotated quite easily by means of rate gyros (Sect. 5.2.5).

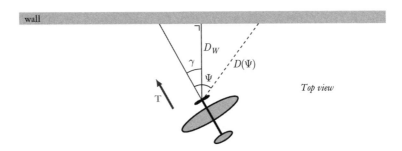

Figure 6.1   A frontal approach toward a flat surface (wall). The distance from the wall D_W is defined as the shortest distance to the wall surface. The approach angle γ is null when the translation $\mathbf{T}$ is perpendicular to the wall. $D(\Psi)$ represents the distance from the wall under a particular azimuth angle Ψ. Note that the drawing is a planar representation and that, in general, D is a function not only of Ψ, but also of the elevation Θ.

We now consider a situation where the robot approaches a wall represented by an infinitely large flat surface, in straight and level flight, at a given angle of approach γ (Fig. 6.1). Note the translation vector points at the center of the FOV. The simplest case is a perpendicular approach to the wall ($\gamma = 0°$). Figure 6.2a displays the OF field that arises in the frontal part of the FOV. This field is divergent, which means that all OF vectors radiate from the focus of expansion (FOE). Note that the amplitude of the OF vectors are not proportional to the sine of the eccentricity α (angle from the FOE), as predicted by equation (5.2). This would be the case only when all the distances $D(\Psi, \Theta)$ from the surface are equal (i.e. a spherical obstacle centered at the location of the vision system). Instead, in the case of a flat surface, the distance increases as the elevation and azimuth angles depart from $0°$. Since $D(\Psi, \Theta)$ is the denominator of the optic flow equation (5.1), smaller OF amplitudes are obtained in the periphery. The locus of the viewing directions corresponding to the maximum OF amplitudes is the solid angle[1], defined by $\alpha = 45°$ [Fernandez Perez de Talens and Ferretti, 1975]. This property is useful when deciding how to orient the OFDs, especially with lightweight robots where vision systems spanning the entire FOV cannot be afforded. It is indeed always interesting to look at regions characterised by large image motion in order to optimise the signal-to-noise ratio, especially when other factors such as low velocity tend to weaken the OF amplitude. In addition, it is evident that looking $90°$ from the forward direction would not help much when it comes to collision avoidance. It is equally important to note that looking straight ahead is useless since this would cause very week and inhomogeneous OF around the FOE.

We now explore what happens when the distance from the surface D_W decreases over time, simulating a robot that actually progresses towards the wall. In Figure 6.2a, third column, the signed[2] OF amplitude p at $\Psi = \pm 45°$ is plotted over time. Both curves are obviously symmetrical and the values are inversely proportional to D_W, as predicted by equation (5.1). Since these signals are asymptotic in $D_W = 0$ m, they constitute good cues for imminent collision detection. For instance, a simple threshold at $|p| = 30°/s$ would suffice to trigger a warning 2 m before collision (see vertical and horizontal dashed lines in Figure 6.2a, on the right). According

---

[1] Because of the spherical coordinates, this does not exactly translate into a circle in elevation-azimuth graphs, i.e. $\alpha \neq \sqrt{\Psi^2 + \Theta^2}$.

[2] Projecting $\mathbf{p}$ on the Ψ axis, rightward OF is positive, whereas leftward OF is negative.

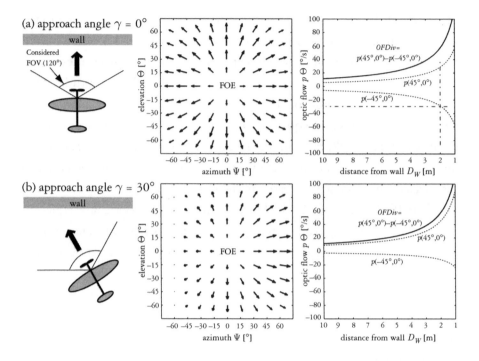

Figure 6.2 Motion fields generated by forward motion at constant speed (2 m/s). (a) A frontal approach toward a wall. (b) An approach at $30°$. The first column depicts the the robot trajectory as well as the considered FOV. The second column shows the motion fields occurring in each situation. The third column shows the signed OF amplitudes p at $\pm45°$ azimuth as a function of the distance from the wall D_W.

to equation (5.1), this distance fluctuates with the airplane velocity $\|\mathbf{T}\|$, but in a favourable manner. Since the optic-flow amplitude is proportional to the translational velocity ($p \sim \|\mathbf{T}\|$), the warning would be triggered earlier (at 3 m instead of 2 m before the wall for a plane fling at 3 m/s instead of 2 m/s), hence permitting a greater distance for an avoidance action. In fact, by using a fixed threshold on the OF, the ratio $\frac{D_W}{\|\mathbf{T}\|}$ is kept constant. This ratio is nothing else than the time to contact (TTC, see Sect. 3.3.3).

Based on these properties, it would be straightforward to place a single OFD directed in a region of maximum OF amplitude (at $\alpha = 45°$) to ensure a good signal-to-noise ratio of the OFD and simply monitor when this value reaches a preset threshold. However, in reality, the walls are not as high as they are wide (Fig. 4.17a), and consequently, OFDs oriented at non null elevation have a higher risk of pointing at the ground or the ceiling. For this reason, the most practical orientation is $\Psi = 45°$ and $\Theta = 0°$.

What happens if the path direction is not perpendicular to the obstacle surface? Figure 6.2(b) depicts a situation where $\gamma = 30°$. The OF amplitude to the left is smaller whereas the amplitude to the right is larger. In this particular case, a possible approach is to sum (or average) the left and right OF amplitudes, which results in the same curve as in the perpendicular approach case (compare the curves labelled *OFDiv*). This sum is proportional to the OF field divergence and is therefore denoted *OFDiv*. This method[3] of detecting imminent collision using a minimum number of OFDs enables the *OFDiv* signal to be measured by summing two symmetrically oriented OFDs, both detecting OF along the equator.

Before testing this method, it is interesting to consider how the OF amplitude behaves on the frontal part of the equator, when the plane approaches the wall at angles varying from between $0°$ and $90°$ and what would be the consequences of the approaching angle on *OFDiv*. This can be worked out using the motion parallax equation (5.2) while replacing α by Ψ since we are only interested in what happens at $\Theta = 0°$. The distance from the obstacle in each viewing direction (see Figure 6.1 for the geometry and notations) is given by:

$$D(\Psi) = \frac{D_W}{\cos(\Psi + \gamma)} . \qquad (6.1)$$

Then, by using motion parallax, the OF amplitude can be calculated as:

$$p(\Psi) = \frac{\|\mathbf{T}\|}{D_W} \sin \Psi \cdot \cos(\Psi + \gamma). \qquad (6.2)$$

Figure 6.3, left column, displays the OF amplitude in every azimuthal direction as well as for a set of approaching angles ranging from $0°$ (perpendicular approach) to $90°$ (parallel to the wall). The second column plots the sum of the left and right sides of the first column graphs. This sum corresponds to *OFDiv* as if it was computed for every possible azimuth in the

---

[3] This way of measuring the OF divergence is reminiscent of the minimalist method proposed by Ancona and Poggio [1993], using Green's theorem [Poggio *et al.*, 1991].

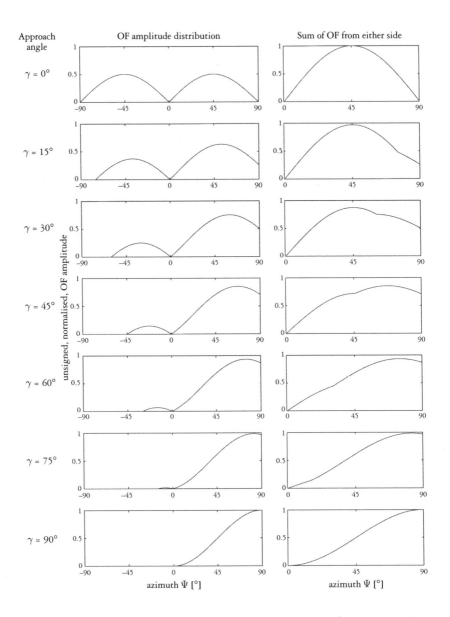

Figure 6.3 A series of graphs displaying the repartition of the unsigned, normalised OF amplitudes on the equator of the vision sensor (i.e. where $\Theta = 0$) in the case of a frontal approach toward a flat surface at various approaching angles γ. The second column represents the symmetrical sum of the left and right OF amplitudes, as if the graphs to the left were folded vertically at $\Psi = 0$ and the OF values for every $|\Psi|$ were summed together. This sum corresponds to *OFDiv* as if it was computed for every possible azimuth in the frontal part of the equator.

frontal part of the equator. Up to $\gamma = 30°$, the sum of OF maintains a maximum at $|\Psi| = 45°$. For wider angles of approach, the peak shifts toward $|\Psi| = 90°$.

Before drawing conclusions concerning optimal OFD viewing directions for estimating *OFDiv*, one should take into consideration the complexity of the avoidance manoeuvre, which essentially depends on the approach angle. When perpendicularly approaching the wall, the airplane must perform at least a 90° turn. Instead, when following an oblique course (e.g. $\gamma = 45°$), a 45° turn in the correct direction is enough to avoid colliding with the wall, and so on until $\gamma = 90°$ where no avoidance action is required at all. For two OF measurements at $\Psi = \pm45°$, the *OFDiv* signal (Fig. 6.3, right column) is at its maximum when the plane approaches perpendicularly and decreases to 70 % at 45°, and to 50 % at 90° (where no action is required). As a result, the imminent collision detector is triggered at a distance 30 % closer to the wall when the approaching angle is 45°. The plane could also fly along the wall ($\gamma = 90°$) without any warning, at a distance 50 % closer to the wall than if it would have had a perpendicular trajectory. Therefore, this strategy for detecting imminent collisions is particularly interesting, since it automatically adapts the occurrence of the warning to the angle of approach and the corresponding complexity of the required avoidance manoeuvre.

A similarly interesting property of the *OFDiv* signal, computed as a sum of left and right OF amplitudes, arises when approaching a corner (Fig. 6.4). Here the minimal avoidance action is even greater than in the

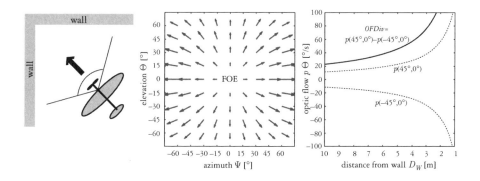

Figure 6.4 Same as Figure 6.2, but for the case of an approach toward a corner.

worst situation with a simple wall since the plane has to turn by more than $90°$ (e.g. $135°$ when approaching on the bisector). Fortunately, the *OFDiv* signal is significantly higher in this case as a result of the average distances from the surrounding walls being smaller (compare *OFDiv* curve in Figure 6.4 and 6.2).

To sum up, two OFDs are theoretically sufficient for detecting imminent collisions. The best way of implementing them on the robot is to orient their viewing directions at $\Psi = \pm 45°$ and $\Theta = 0°$ and to place them horizontally in order to detect radial OF along the equator. Summing their outputs creates an *OFDiv* signal that can be used with a simple threshold for detecting impending collisions. A further interesting property of this signal is that it reaches the same threshold at slightly different distances from the obstacles, as well as the way this varying distance is adapted (i) to the complexity of the minimal required avoidance action (i.e. required turning angle), and (ii) to the flight velocity. We now know how to detect imminent collision in theory, but we still need to design an actual controller to steer the robot.

6.1.2 Control Strategy

The steering control strategy we propose is largely inspired by the recent study of Tammero and Dickinson [2002a] on the behaviour of free-flying fruitflies (see also Sect. 3.4.3). They showed that:

- OF divergence experienced during straight flight sequences is responsible for triggering saccades,
- the direction of the saccades (left or right) is the opposite with regard to the side experiencing larger OF, and
- during saccades, no visual feedback seems to be used.

The proposed steering strategy can thus be divided into two mechanisms: (i) maintaining a straight course and (ii) turning as quickly as possible as soon as an imminent collision is detected.

Course Stabilisation

Maintaining a straight course is interesting in two respects. On the one hand, it spares energy in flight since a plane that banks must produce additional lift in order to compensate for the centrifugal force. On the other hand, it provides better conditions for estimating OF since the airplane is in level flight and the frontal OFDs will see only the textured walls and thus not the ceiling and floor of the test arena (Fig. 4.17a).

In Section 3.4.2, we mentioned that flying insects are believed to implement course stabilisation using both visual and vestibular cues. In order to achieve straight course with our artificial systems, we propose to rely exclusively on gyroscopic data. It is likely that the artificial rate gyro has a higher accuracy than the halteres' system, especially at low rotation rates. Moreover, decoupling the sensory modalities by attributing the rate gyro to the course stabilisation and the vision to collision avoidance simplifies the control structure. With an airplane, course stabilisation can thus be easily implemented by means of a proportional feedback loop connecting the rate gyro to the rudder servomotor. Note that, unlike the plane, the *Khepera* does not need a gyro for moving in a straight line since its wheel speeds are regulated and almost no slipping occurs between the wheels and the ground. Thus, no active course stabilisation mechanism is required.

Collision Avoidance

Saccades (quick turning actions) represent a means of avoiding collisions. To detect imminent collisions, we propose to rely on the spatio-temporal integration of motion (STIM) model (Sect. 3.3.3), which spatially and temporally integrates optic flow from the left and right eyes. Note that, according to Tammero and Dickinson [2002b], the STIM model remains the one that best explains the landing and collision-avoidance responses in their experiments. Considering this model from an engineering viewpoint, imminent collision can be detected during straight motion using the *OFDiv* signal obtained by summing left and right OF amplitudes measured at $\pm 45°$ azimuth (Sect. 6.1.1). Therefore, two OFDs must be mounted horizontally and oriented at $45°$ off the longitudinal axis of the robot. Let us denote the output signal of the left detector *LOFD* and that of the right one *ROFD*. *OFDiv* is thus obtained as follows:

$$OFDiv = ROFD + (-LOFD). \tag{6.3}$$

Note that OFD output signals are signed OF amplitudes that are positive for rightward motion. In order to prevent noisy transient OFD signals (that may occur long before an actual imminent collision occurs) from triggering a saccade, the *OFDiv* signal is low-pass filtered. Figure 6.5 outlines the comparison between the fly model and the system proposed as the robot control strategy. Note that a leaky integrator (equivalent to a low-pass filter) is also present in the fly model and accounts for the fact that weak motion stimuli do not elicit any response [Borst, 1990].[4]

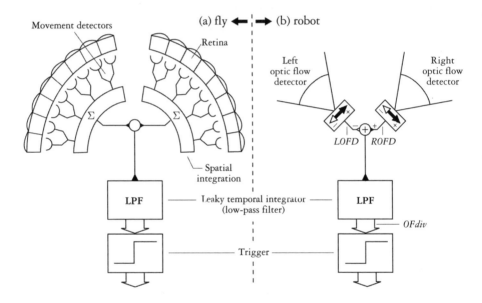

Figure 6.5 The STIM model (to the left, adapted from Borst and Bahde, 1988) as compared to the system proposed for our robots (to the right). (a) The output of motion detectors (EMDs) sensitive to front-to-back motion are spatially pooled from each side. The resulting signal is then fed into a leaky temporal integrator (functionally equivalent to a low-pass filter). When the temporal integrator reaches a threshold, a preprogrammed motor sequence can be performed, either to extend legs or to trigger a saccade (see Sect. 3.3.3 for further discussion). (b) The system proposed for imminent collision detection in our robots is very similar. The spatial pooling of EMDs on the left and right regions of the field of view are simply replaced by two OFDs.

---

[4] However, the time constant of the low-pass filter could not be precisely determined.

As pointed out in Section 6.1.1, the output signal *OFDiv* reaches the threshold in a way that depends on the speed, the angle of approach and the geometry of the obstacle. For instance, the higher the approaching speed, the earlier the trigger will occur.

Turning Direction

As seen in Chapter 5, close objects generate larger translational optic flows. The left-right asymmetry between OFD outputs prior to each saccade can thus be used in order to decide the direction of the saccade. The same strategy seems to be used by flies to decide whether to turn left or right [Tammero and Dickinson, 2002a]. A new signal is thus defined, which measures the difference between left and right absolute OF values:

$$OFDiff = |ROFD| - |LOFD| . \qquad (6.4)$$

A closer obstacle to the right results in a positive *OFDiff*, whereas a closer obstacle to the left produces a negative *OFDiff*.

Finally, Figure 6.6 shows the overall signal flow diagram for saccade initiation and direction selection. Note that *OFDiv*, as computed in equation (6.3), is not sensitive to yaw rotation since the rotational component is detected equally by the two OFDs, whose outputs are subtracted.[5] Unlike *OFDiv*, *OFDiff* does suffer from RotOF and must be corrected for this using the rate gyro signal.

The global control strategy encompassing the two mechanisms of course stabilisation and collision avoidance (Fig. 6.7) can be organised into a subsumption architecture [Brooks, 1999].

6.1.3   Results on Wheels

The steering control proposed in the previous Section (without course stabilisation) was first tested on the *Khepera* robot in a square arena (Fig. 4.15b).

---

[5] A property also pointed out by Ancona and Poggio [1993]. This method for estimating flow divergence is independent of the location of the focus of expansion. In our case, this means that the measured divergence remains unaltered when the FOE shifts left and right due to rotation.

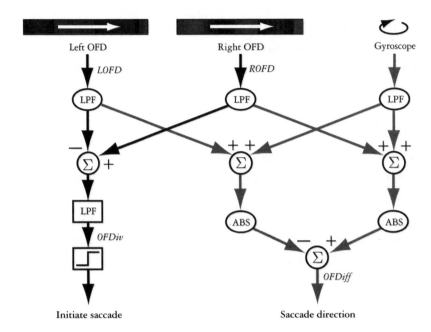

Figure 6.6 A signal flow diagram for saccade initiation (collision avoidance) based on horizontal OF divergence and rotation rate as detected by the yaw rate gyro. The arrows at the top of the diagram indicate the positive directions of OFDs and rate gyro. LPF stands for low-pass filter and ABS is the absolute value operator. The signals from the OFDs and rate gyro are first low-pass filtered to cancel out high-frequency noise (Fig. 5.8). Below this first-stage filtering, one can recognise, to the left (black arrows), the STIM model responsible for saccade initiation and, to the right (grey arrows), the pathway responsible for deciding whether to turn left or right.

The robot was equipped with its frontal camera (Fig. 4.1), and two OFDs with FOVs of 30° were implemented using 50 % of the available pixels (Fig. 6.8). The *OFDiv* signal was computed by subtracting the output of the left OFD from the output of the right OFD (see equation 6.3).

As suggested above, the steering control was composed of two states: (i) straight, forward motion at constant speed (10 cm/s) during which the system continuously computed *OFDiv*, (ii) rotation for a fixed amount of time (1 s) during which sensory information was discarded. A period of one second was chosen in order to produce a rotation of approximately 90°, which is in accordance with what was observed by Tammero and Dickinson [2002a]. A transition from state (i) to state (ii) was triggered whenever

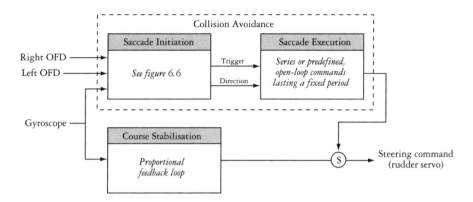

Figure 6.7 The proposed steering strategy. To the left are the sensory inputs (optic-flow detectors and rate gyro) and to the right is the control output (steering command). The encircled S represents a suppressive node; in other words, when active, the signal coming from above replaces the signal usually going horizontally trough the node.

OFDiv reached a threshold whose value was experimentally determined beforehand. The direction of the saccade was determined by the asymmetry *OFDiff* between left and right OFDs, i.e. the *Khepera* turned away from the side experiencing the larger OF value.

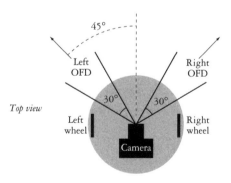

Figure 6.8 The arrangement of the OFDs on the *Khepera* equipped with the frontal camera (see also Fig. 4.11a) for the collision avoidance experiment.

By using this control strategy, the *Khepera* was able to navigate without collisions for more than 45 min (60'000 sensory-motor cycles), during which time it was engaged in straight motion 84% of the time, spending

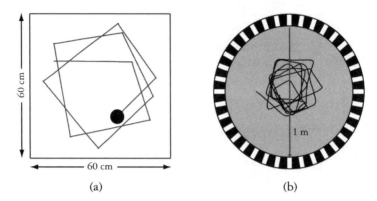

Figure 6.9 (a) Collision avoidance with the *Khepera*. The path of the robot in autonomous steering mode: straight motion with saccadic turning actions whenever image expansion (*OFDiv*) reached a predefined threshold. The black circle represents the *Khepera* at its starting position. The path has been reconstructed from wheel encoders. (b) For comparison, a sample trajectory (17 s) within a textured background of a real fly Drosophila melanogaster [Tammero and Dickinson, 2002a].

only 16% of the time in saccades. Figure 6.9 shows a typical trajectory of the robot during this experiment and highlights the resemblance with the flight behaviour of flies.

6.1.4   Results in the Air

Encouraged by these results, we proceeded to autonomous steering experiments with the *F2* (Sect. 4.1.3) in the arena depicted in Figure 4.17(a). The 30-gram airplane was equipped with two miniature cameras oriented 45° off the forward direction, each providing 28 pixels for the left and right OFDs spanning 40° (Fig. 6.10).

A radio connection (Sect. 4.2.3) with a laptop computer was used in order to log sensor data in real-time while the robot was operating. The plane was started manually from the ground by means of a joystick connected to a laptop. When it reached an altitude of approximately 2 m, a command was sent to the robot to switch it into autonomous mode. While in this mode, the human pilot had no access to the rudder (the vertical control surface), but could modify the pitch angle by means of the elevator (the horizontal

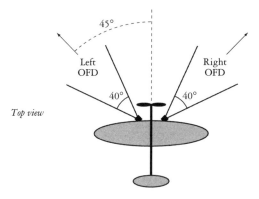

Figure 6.10 The arrangement of the two OFDs on the *F2* airplane. See also the picture in Figure 4.11(b).

control surface).[6] The sensory-motor cycle typically lasted 80 ms. During this period, data from on-board sensors were processed, commands for the control surfaces were issued, and significant variables were sent to the laptop for off-line analysis. About 50% of this sensory-motor cycle was spent in wireless communication, which means that control robustness could be further improved by shortening the sensory-motor period if no data needed to be sent to the ground station.

During saccades, with time lengths set to 1 s[7], the motor was set to full power, the rudder deflection followed an experimentally optimised curve up to full deflection, and the elevator was slightly pulled to compensate for the decrease in lift during banked turns. At the end of a saccade, the plane was programmed to resume straight flight while it was still banked. Since banking always produces a yaw movement, the proportional controller based on the yaw rate gyro (Sect. 6.1.2) compensated for the inclination and forced the plane back to level flight. We also implemented an inhibition period after the saccade, during which no other turning actions could be triggered. This allowed for the plane to recover straight flight be-

---

[6] If required, the operator could switch back to manual mode at any moment, although a crash into the curtained walls of the arena did not usually damage the lightweight airplane.

[7] This time length was chosen in order to produce roughly 90° turns per saccade. However, this angle could fluctuate slightly depending on the velocity that the robot displayed at the saccade start.

fore deciding whether to perform another saccade. In our case, the inhibition was active as long as the rate gyro indicated an absolute yaw rotation larger than $20°/s$. This inhibition period also permitted a resetting of the *OFDiv* and *OFDiff* signals that could be affected by the strong optic-flow values occurring just before and during the saccade due to the nearness of the wall.

Before testing the airplane in autonomous mode, the *OFDiv* threshold for initiating a saccade (Fig. 6.6) was experimentally determined by flying manually in the arena and recording OFD signals while frontally approaching a wall and performing an emergency turn at the last possible moment. The recorded OFD data was analysed and the threshold was chosen on the basis of the value reached by *OFDiv* just before the avoidance action.

An endurance test was then performed in autonomous mode. The *F2* was able to fly without collision in the 16×16 m arena for more than 4 min without any steering intervention.[8] The plane was engaged in saccades only 20% of the time, thus indicating that it was able to fly in straight trajectories except when very close to a wall. During the 4 min, it generated 50 saccades, and covered approximately 300 m in straight motion.

Unlike the *Khepera*, the *F2* had no embedded sensors allowing for a plotting of its trajectory. Instead, Figure 6.11 displays a detailed 18-s sample of the data acquired during typical autonomous flight. Saccade periods are highlighted with vertical gray bars spanning all the graphs. In the first row, the rate gyro output provides a good indication of the behaviour of the plane: straight trajectories interspersed with turning actions, during which the plane could reach turning rates up to $100°/s$. OF was estimated by the OFDs computed from the 1D images shown in the second row. The miniature cameras did not provide very good image quality. As a result, OFD signals were not always very accurate, especially when the plane was close to the walls (few visible stripes) and had a high rotational velocity. This situation happened most often during the saccade inhibition period. Therefore, we decided to clamp *OFDiv* and *OFDiff* (two last rows of Figure 6.11) to zero whenever the rate gyro was above $20°/s$.

---

[8] Video clips showing the behaviour of the plane can be downloaded from http://book.zuff.info

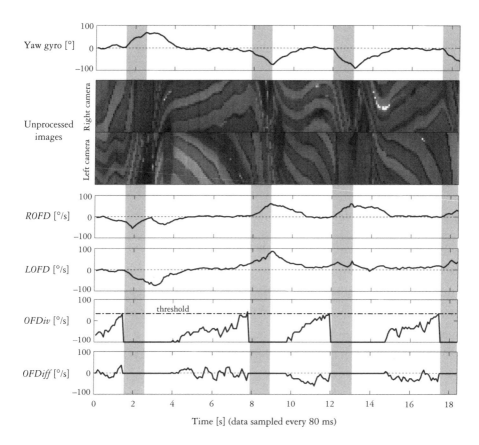

Figure 6.11   The sensor and OF data during autonomous flight (approximately 18 s are displayed). The first row represents the yaw rate gyro indicating how much the plane was rotating (rightward positive). The second row displays the raw images as seen by the two cameras every sensory-motor cycle. Only the 28 pixels used for OF detection are displayed for each camera. The third and fourth rows are the OF as estimated by the left and the right OFDs, respectively. The fifth and sixth rows show the OF divergence *OFDiv* and difference *OFDiff* when the absolute value of the rate gyro was below $20°/s$, i.e. when the plane was flying almost straight. The dashed horizontal line in the *OFDiv* graph represents the threshold for triggering a saccade. The gray vertical lines spanning all the graphs indicate the saccades themselves. The first saccade was leftward and the next three were rightward, as indicated by the rate gyro values in the first row. Adapted from [Zufferey and Floreano, 2006].

When *OFDiv* reached the threshold indicated by the dashed line, a saccade was triggered. The direction of the saccade was based on *OFDiff* and is plotted in the right-most graph. The first turning action was leftward since *OFDiff* was positive when the saccade was triggered. The remaining turns were rightward because of the negative values of the *OFDiff* signal. When the approach angle was not perpendicular, the sign of *OFDiff* was unambiguous, as in the case of the third saccade. In other cases, such as before the second saccade, *OFDiff* oscillated around zero because the approach was almost perfectly frontal. Note however that in such cases, the direction of the turning action was less important since the situation was symmetrical and there was no preferred direction for avoiding the wall.

6.1.5   Discussion

This first experiment in the air showed that the approach of taking inspiration from flies can enable a reasonably robust autonomous steering of a small airplane in a confined arena. The control strategy of using a series of straight sequences interspersed with rapid turning actions was directly inspired by the flies' behaviour (Sect. 3.4.3). While in flies some saccades are spontaneously generated in the absence of any visual input, reconstruction of OF patterns based on flies' motion through an artificial visual landscape suggested that image expansion plays an fundamental role in triggering saccades [Tammero and Dickinson, 2002a]. In addition to providing a means of minimising rotational optic flow, straight flight sequences also increase the quality of visual input by maintaining the plane horizontal. In our case, the entire saccade was performed without sensory feedback. During saccades, biological EMDs are known to operate beyond their linear range where the signal could even be reversed because of temporal aliasing [Srinivasan *et al.*, 1999]. However, the role of visual feedback in the control of these fast turning manoeuvres is still under investigation [Tammero and Dickinson, 2002b]. Halteres' feedback is more likely to have a major impact on the saccade duration [Dickinson, 1999]. Although the *F2* did not rely on any sensory feedback during saccade, the use of gyroscopic information could provide an interesting way of controlling the angle of the rotation. Finally, the precise roles of halteres and vision in course (or gaze) stabilisation of flies is still unclear (Sect. 3.4.2). Both sensory modalities are believed to have an influence, whereas in the *F2*, course stabilisation and OF derotation (which can be seen as the placeholder of gaze stabilisation in flies) rely exclusively on gyroscopic information.

More recently, a similar experiment has been reproduced with the 10-gram *MC2* airplane in a much smaller arena [Zufferey *et al.*, 2007]. Since the *MC2* is equipped with an anemometer, it had the additional benefit of autonomously controlling its airspeed. However, autonomous steering did not correspond to complete autonomous operation since the elevator sill needed to be remotely operated by a human pilot whose tasks consisted in maintaining reasonable altitude above the ground. Therefore, we will now explore how also the altitude control could be automated.

6.2   Altitude Control

Now that lateral steering has been solved, altitude control is the next step. For the sake of simplicity, we assume straight flight (and thus a zero roll angle) over a flat surface. Only the pitch angle is let free to vary in order to act on the altitude. This simplification is reasonably representative of what happens between the saccades provoked by the steering controller proposed above. The underlying motivation is that if these phases of straight motion are long enough and the saccade periods are short enough, it may be sufficient to control altitude only during straight flight, when the plane is level. This would simplify the altitude control strategy while ensuring that ventral cameras are always oriented towards the ground.

6.2.1   Analysis of Ventral Optic Flow Patterns

The situation of interest is represented by an aircraft flying over a flat surface (Fig. 6.12) with a camera pointing downwards. The typical OF pattern that occurs in the bottom part of the FOV is simpler than that taking place in frontal approach situations. All OF vectors are oriented in the same direction, from front to back. According to equation (5.1), their amplitude is inversely proportional to the distance from the ground ($p \sim \frac{1}{D(\Psi,\Theta)}$). The maximum OF amplitude in the case of level flight (zero pitch) is located at $\Theta = -90°$ and $\Psi = 0°$. Therefore, a single OFD pointing in this direction (vertically downward) could be a good solution for estimating altitude since its output is proportional to $\frac{1}{D_A}$.

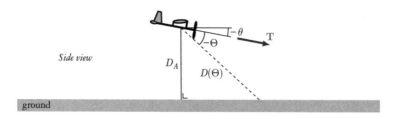

Figure 6.12 An airplane flying over a flat surface (ground). The distance from the ground D_A (altitude) is defined as the shortest distance (perpendicular to the ground surface). The pitch angle θ is null when **T** is parallel to the ground. $D(\Theta)$ represents the distance from the ground at a certain elevation angle Θ in the visual sensor reference frame. Note that the drawing is a 2D representation and that D is generally a function not only of Θ, but also of the azimuth Ψ.

Let us now restrict the problem to 2D and analyse what happens to the 1D OF field along $\Psi = 0°$ when the airplane varies its pitch angle in order to change its altitude. As before, the motion parallax equation (5.2) permits better insight into this problem:

$$D(\Theta) = \frac{D_A}{-\sin(\Theta + \theta)} \implies p(\Theta) = \frac{\|\mathbf{T}\|}{D_A}\sin\Theta \cdot \sin(\Theta + \theta). \quad (6.5)$$

Based on this equation, Figure 6.13 shows the OF amplitude as a function of the elevation for various cases of negative pitch angles. Of course, the situation is symmetrical for positive pitch angles. These graphs reveal that the location of the maximum OF is $\Theta = -90°$ plus half the pitch angle. For example, if $\theta = -30°$, the peak is located at $\Theta = -90 - 30/2 = -75°$ (see the vertical dashed line in the third graph). This property can be derived mathematically from equation (6.5):

$$\frac{dp}{d\Theta} = \frac{\|\mathbf{T}\|}{D_A}\sin(2\Theta + \theta) \quad \text{and} \quad \frac{dp}{d\Theta} = 0 \iff \Theta_{\max} = \frac{\theta + k\pi}{2}. \quad (6.6)$$

As seen in Figure 6.13, the peak amplitude weakens only slightly when the pitch angle departs from $0°$. Therefore, a single OFD, pointing vertically downward, is likely to provide sufficient information to control the altitude, especially for an airplane rarely exceeding a $\pm10°$ pitch angle.

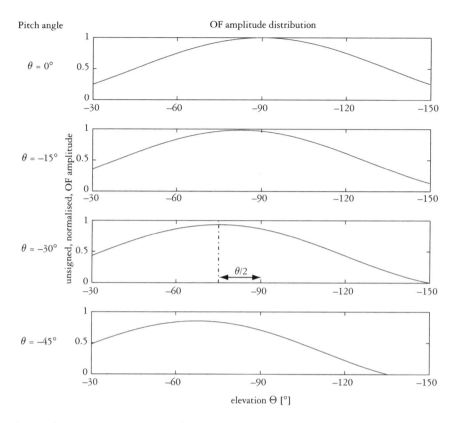

Figure 6.13 The repartition of the unsigned, normalised OF amplitudes in the longitudinal direction (i.e. $\Psi = 0°$) in the case of flight over a flat surface at various pitch angles θ.

6.2.2 Control Strategy

As suggested in Section 3.4.4, altitude can be controlled by maintaining the ventral optic flow constant. This idea is based on experiments with honey-bees that seem to use such a mechanism for tasks like grazing landing and control of flight speed. As long as the pitch angle is small (typically within $\pm 10°$), it is reasonable to use only one vertical OFD. For larger pitch angles, it is worth tracking the peak OF value. In this case, several OFDs pointing in various directions (elevation angles) must be implemented and only the OFD producing the maximum output (whose value is directly related to the altitude) is taken into account in the control loop (winner-take-all).[9]

---

[9] A similar strategy has been used to provide an estimate of the pitch angle with respect to a flat ground [Beyeler *et al.*, 2006].

The control loop linking the ventral OF amplitude to the elevator should integrate a derivative term in order to dampen the oscillation that may arise due to the double integrative effect existing between the elevator angle and the variation of the altitude. We indeed have $\frac{dD_A}{dt} \sim \theta$ (Fig. 6.12) and $\frac{d\theta}{dt}$ is roughly proportional to the elevator deflection.

6.2.3   Results on Wheels

In order to assess the suggested altitude control strategy, we implemented it as a wall-following mechanism on the *Khepera* with the camera oriented laterally (Fig. 6.14). In this situation, the distance from the wall corresponds to the altitude of the aircraft and the rotation speed of the *Khepera* around its yaw axis is comparable to the effect of the elevator deflection command on an airplane. Since the wheeled robot is not limited with regard to the orientation angle it can take with respect to the wall, we opted for the strategy with several OFDs sensitive to longitudinal OF. Therefore, four adjacent OFDs were implemented, each using a subpart of the pixels of the single 1D camera mounted on the *kevopic* board.

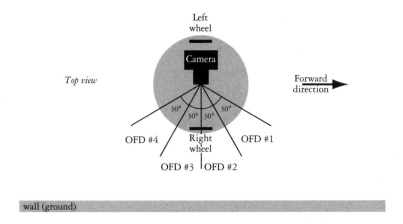

Figure 6.14   An outline of the *Khepera* equipped with the wide FOV lateral camera (see also Figure 4.11a) for the wall-following experiment. Four OFDs were implemented, each using a subpart of the pixels.

A proportional-derivative controller attempts maintaining the OF amplitude constant by acting on the differential speed between the left and right wheels. As previously, the yaw rate gyro signal is used to derotate

OF. The OFD value employed by the controller is always the one producing the highest output among the four OFDs. In practice, only the two central OFDs are the ones often used, but the external ones can also be used when the *Khepera* takes very steep angles with respect to the wall.

Several tests were performed with a 120-cm-long wall (Fig. 6.15). Although the robot did not always keep the same distance from the wall, the tests showed that such a simple control strategy based on optic flow could produce a reliable altitude control. Note that this would not have been possible without careful derotation of OF.

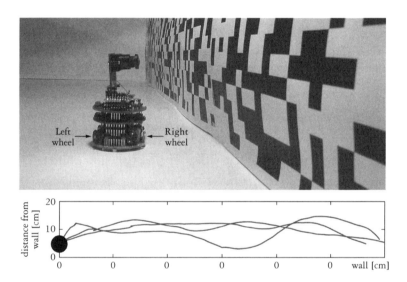

Figure 6.15 Altitude control (implemented as wall following) with the *Khepera*. Top: The 120-cm long setup and the *Khepera* with the lateral camera. Bottom: Wall following results (3 trials). The black circle indicates the robot's initial position. Trajectories are reconstructed from wheel encoders.

6.2.4   Discussion

The presented control strategy relies on no other sensors than vision and a MEMS gyroscope, and is therefore a good candidate for ultra-light flying robots. Furthermore, this approach to optic-flow-based altitude control proposes two new ideas with respect to the previous work [Barrows *et al.*, 2001; Chahl *et al.*, 2004; Ruffier and Franceschini, 2004] presented in Section 2.2. The first is the pitching rotational optic-flow cancellation using the rate gyro, which allows the elimination of the spurious signals occurring whenever a pitch correction occurs. The second is the automatic tracking of the ground perpendicular distance, removing the need for measuring the pitch angle with another sensor. Although much work remains to be done to validate this approach[10], these advantages are remarkable since no easy solution exists outside vision to provide a vertical reference to ultra-light aircraft.

However, the assumption made at the beginning of this Section holds true only for quite specific cases. The fact that the airplane should be flying straight and over flat ground most of the time is not always realistic. The smaller the environment, the higher is the need for frequent turning actions. In such situations, the ventral sensor will not always be pointing vertically at the ground. For instance, with the *MC2* flying in its test arena (Fig. 4.17b), the ventral camera is often pointing at the walls as opposed to the ground, thus rendering the proposed altitude control strategy unusable.

Therefore, we propose in the next Section a different approach in order to finally obtain a fully autonomous flight. The underlying idea is to eliminate the engineering tendency of reducing collision avoidance to 2D sub-problems and then assume that combining the obtained solutions will resolve the original 3D problem. Note that this tendency is often involuntarily suggested by biologists who also tend to propose 2D models in order to simplify experiments and analysis of flight control in insects.

6.3   3D Collision Avoidance

Controlling heading and altitude separately resembles the airliners way of flying. However, airliners generally fly in an open space and need to maintain a level flight in order to ease traffic control. Flying in confined areas is closer to an aerobatic way of piloting where the airplane must constantly roll and pitch in order to avoid collisions. Instead of decoupling

---

[10] Testing various arrangements of optic-flow detectors with or without overlapping field-of-views, or explicitly using the information concerning the pitch angle within the control loop. Note that this method would likely require a quite high resolution of the optic-flow field and thus a high spatial frequency on the ground as well as a number of optic-flow detectors.

lateral collision avoidance and vertical altitude control, we here propose to think in terms of 3D collision avoidance. Finally, the primary goal is not to fly level and as straight as possible, but rather to avoid any collisions while remaining airborne. To do so, we propose to return to the seminal thoughts by Braitenberg [1984] and to think in terms of direct connections between the various OFDs and the airplane controls.

6.3.1   Optic Flow Detectors as Proximity Sensors

In order to apply Braitenberg's approach to collision avoidance, the OFDs need to be turned into proximity sensors. According to equation (5.3), this is possible only if they are

- carefully derotated (Sect. 5.2.5),
- radially oriented with respect to the FOE,
- pointed at a constant eccentricity α (Figs 5.1 and 5.4).

Since the *MC2* is equipped with two cameras, one horizontal pointing forward and a second one pointing downwards, one can easily design three OFDs following this policy. Figure 6.16 shows the regions covered by the two cameras. If only the gray zones are chosen, the resulting OFDs are effectively oriented radially and at a fixed eccentricity of $45°$. Note that this angle is not only chosen because it fits the available cameras, but also because the maximum OF values occur at $\alpha = 45°$ (Sect. 6.1.1). As a result, the *MC2* is readily equipped with three proximity sensors oriented at $45°$ from the moving direction, one to the left, one to the right, and one in the ventral region. A forth OFD could have been located in the top region (also at $45°$ eccentricity), but since the airplane never flies inverted (due to the passive stability) and the gravity attracts it towards the ground, there is no need for sensing obstacles in this region. In addition, the ceiling of the test arena (Fig. 4.17b) is not equipped with visual textures that could be accurately detected by an OFD.

In this new approach, great care must be taken to carefully derotate the OFDs, otherwise their signals may be overwhelmed by spurious rotational OF and no longer representative of proximity. In order to achieve a reasonable signal-to-noise ratio, both the rate gyro and the OF signals are low-pass filtered using a first-order filter prior to the derotation process (Sect. 5.2.5).

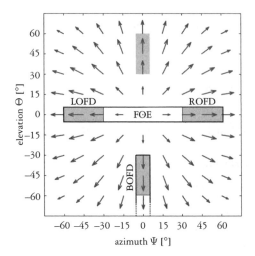

Figure 6.16 An azimuth-elevation graph displaying the zones (thick rectangles) covered by the cameras mounted on the *MC2* (see also Figure 4.11c). By carefully defining the sub-regions where the I2A is applied (gray zones within the thick rectangles), three radial OFDs can be implemented at an equal eccentricity of 45° with respect to the focus of expansion (FOE). These are prefixed with L, B, and R for left, bottom and right, respectively.

Such a low-pass filtered, derotated OFD is from hereon denoted DOFD. In addition to being derotated, a DOFD is unsigned (i.e. only positive) since only the radial OF is of interest when indicating proximity. In practice, if the resulting OF, after filtering and derotation, is oriented towards the FOE and not expanding from it, the output of the DOFD is clamped to zero.

6.3.2 Control Strategy

Equipped with such DOFDs that act as proximity sensors, the control strategy becomes straightforward. If an obstacle is detected to the right (left), the airplane should steer left (right) using its rudder. If the proximity signal increases in the ventral part of the FOV, the airplane should steer up using its elevator. This is achieved through direct connections between the DOFDs and the control surfaces (Fig. 6.17). A transfer function Ω is employed on certain links to tune the resulting behaviour. In practice, simple multiplicative factors (single parameter) or combinations of a threshold and a factor (two parameters) have been found to work fine.

In order to maintain airspeed in a reasonable range (above stall and below over-speed), the anemometer signal is compared to a given set-point before being used to proportionally drive the propeller motor. Note that this air speed control process also ensures a reasonably constant $\|\mathbf{T}\|$ in equation (5.3).

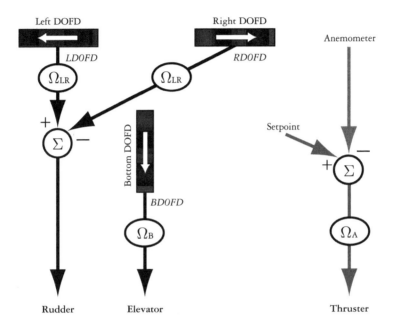

Figure 6.17 A control scheme for completely autonomous navigation with 3D collision avoidance. The three OFDs are prefixed with D to indicate that they are filtered and derotated (this process is not explicitly shown in the diagram). The signals produced by the left and right DOFDs, i.e. *LDOFD* and *RDOFD*, are basically subtracted to control the rudder, whereas the signal from the bottom DOFD, i.e. *BDOFD*, directly drives the elevator. The anemometer is compared to a given set-point to output a signal that is used to proportionally drive the thruster. The Ω-ellipses indicate that a transfer function is used to tune the resulting behaviour. These are usually simple multiplicative factors or combinations of a threshold and a factor.

6.3.3 Results in the Air

The *MC2* was equipped with the control strategy drafted in Figure 6.17. After some tuning of the parameters included in the Ω transfer functions, the airplane could be launched by hand in the air and fly completely au-

tonomously in its arena (Fig. 4.17b)[11]. Several trials were carried out with
the same control strategy and the *MC2* demonstrated a reasonably good robustness.

Figure 6.18 shows data recorded during such a flight over a 90-s period.
In the first row, the higher *RDOFD* signal suggests that the airplane was
launched closer to a wall on its right, which produced a leftward reaction
(indicated by the negative yaw gyro signal) that was maintained throughout
the trial duration. Note that in this environment, there is no good reason
for modifying the initial turning direction since flying in circles close to the
walls is more efficient than describing eights, for instance. However, this
first graph clearly shows that the controller does not simply hold a constant
turning rate. Rather, the rudder deflection is continuously adapted based
on the DOFD signals, which leads to a continuously varying yaw rotation
rate. The average turning rate of approximately 80°/s indicates that a full
rotation is accomplished every 4-5 s. Therefore, a 90-s trial corresponds to
approximately 20 circumnavigations of the test arena.

The second graph shows that the rudder actively reacts to the *BDOFD*
signal, thus continuously affecting the pitch rate. The non-null mean of the
pitch gyro signal is due to the fact that the airplane is banked during turns.
Therefore the pitch rate gyro also measures a component of the overall
circling behaviour. It is interesting to realise that the elevator actions are
not only due to the proximity of the ground, but also of the walls. Indeed,
when the airplane feels the nearness of a wall to its right by means of its
RDOFD, the rudder action increases its leftward bank angle. In this case the
bottom DOFD is oriented directly towards the close-by wall and no longer
towards the ground. In most cases, this would result in a quick increase
in *BDOFD* and thus trigger a pulling action of the elevator. This reaction
is highly desirable since the absence of a pulling action at high bank angle
would result in an immediate loss of altitude.

The bottom graph shows that the motor power is continuously adapted
according to the anemometer value. In fact, as soon as the controller steers
up due to a high ventral optic flow, the airspeed quickly drops, which needs
to be counteracted by a prompt increase in power.

---

[11] A video of this experiment is available for download at http://book.zuff.info

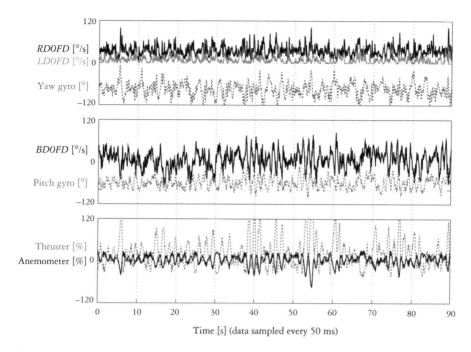

Figure 6.18 A 90-s autonomous flight with the *MC2* in the test arena shown in Figure 4.17(b). The first row shows lateral OF signals together with the yaw rate gyro. The second row plots the ventral OF signal together with the pitch gate gyro. The third graph displays the evolution of the anemometer value together with the motor setting. Flight data are sampled every 50 ms, corresponding to the sensory-motor cycle duration.

6.3.4 Discussion

This Section proposed a 3D adaptation of the Braitenberg approach to collision avoidance by use of optic flow. Braitenberg-like controllers have been widely used on wheeled robots equipped with proximity sensors (see for instance Nolfi and Floreano, 2000). When using optic flow instead of infrared or other kinds of proximity or distance sensors, a few constraints arise. The robot must be assumed to have a stationary translation vector with respect to its vision system. This ensures that $\sin(\alpha)$ in equation (5.3) can be assumed constant. In practice, all airplanes experience some side slip and varying angles of attack, thus causing a shift of the FOE around its longitudinal axis. However, these variations are usually below $10°$ or so and do not significantly affect the use of DOFDs as proximity indicators. Another constraint is that the DOFDs cannot be directed exactly in the frontal direction (null eccentricity) since the translational optic flow would be zero in that region. This means that a small object appearing in the exact center of the FOV can remain undetected. In practice, if the airplane is continuously steering, such small objects quickly shift towards more peripheral regions where they are sensed. Another solution to solve this problem has been proposed by Pichon *et al.* [1990], which consists in covering the frontal blind zone around the FOE by off-centered OFDs. In the case of an airplane, these could be located on the wing leading edge, for example.

A limitation of Braitenberg-like control is its sensitivity to so-called local minima. These occur when two contradicting proximity sensors are active simultaneously at approximately the same level resulting in an oscillatory behaviour that can eventually lead to collision. With an airplane, such a situation typically occurs when a surface is approached perpendicularly. Both left and right DOFDs would output the same value resulting in a rudder command close to zero. Since an airplane cannot slow-down or stop (as would be the case with a wheeled robot) the crash is inevitable unless this situation is detected and handled accordingly. One option to do so is to integrate the solution developed in Section 6.1, i.e. to monitor the global amount of expanding OF and generate a saccade whenever a threshold is reached. Note that saccades can be used both for lateral and vertical steering [Beyeler *et al.*, 2007]. However, the ability to steer smoothly and proportionally to the DOFDs most of the time is highly favourable in case of elongated environments such as corridors or canyons.

Finally, the proposed approach using direct connections between derotated OFDs and control surfaces has proven efficient and implementable on an ultra-light platform weighing a mere 10 g. The resulting behaviour is much more dynamic than the one previously obtained with the *F2* and no strong assumption, such as flat ground or straight motion, was necessary in order to control the altitude. Although the continuously varying pitch and roll angles yielded very noisy OF (because of the aperture problem)[12], the

---

[12] The aperture problem is even worse with the checkerboard patterns used in the *MC2* test arena than with the vertical stripes previously used with the *F2*. This is a result of pitch and roll movements of the airplane dramatically changing the visual content from one image acquisition to the next in the I2A process.

simplicity of the control strategy and the natural inertia of the airplane acting as a low-pass filter produced a reasonably robust and stable behaviour.

6.4   Conclusion

Bio-inspired, vision-based control strategies for autonomous steering and
altitude control have been developed and assessed on wheeled and flying
robots. Information processing and navigation control were performed entirely on the small embedded microcontroller. In comparison to most previous studies in bio-inspired vision-based collision avoidance (see Sect. 2.2),
our approach relied on less powerful processors and lower-resolution visual
sensors in order to enable operation in self-contained, ultra-light robots in
real-time. In contrast to the optic-flow-based airplanes of Barrows *et al.*
[2001] and Green *et al.* [2004] (see also Section 2.2), we demonstrated continuous steering over extended periods of time with robots that were able
to avoid both frontal, lateral and ventral collisions.

The perceptive organs of flying insects have been our main source of inspiration in the selection of sensors for the robots. Although flies possess a
wide range of sensors, the eyes, halteres and hairs are usually recognized as
the most important for flight control (Sect. 3.2). It is remarkable that, unlike most classical autonomous robots, flying insects possess no active distance sensors such as sonars or lasers. This is probably because of the inherent complexity and energy consumption of such sensors. The rate gyro
equipping our robots can be seen as a close copy of the Diptera's halteres
(Sect. 3.2.2). The selected artificial vision system (Sect. 4.2.2) shares with
its biological counterpart an amazingly low resolution. Its inter-pixel angle
(1.4-2.6°) is in the same order of magnitude as the interommatidial angle
of most flying insects (1-5°, see Sect. 3.2.1). On the other hand, the field
of view of our robots is much smaller than that of most flying insects. This
discrepancy is mainly due to the lack of technology allowing for building
miniature, omnidirectional visual sensors sufficiently light to fit the constraints of our microflyers. In particular, little industrial interest exists so far
in the development of artificial compound eyes, and omnidirectional mir-

rors tend to be too heavy. We have partly compensated the lack of omnidirectional vision sensors by using several small vision sensors pointing in the directions of interest. These directions were identified based on the analysis of optic-flow patterns arising in specific situations. We demonstrated that three 1D optic-flow detectors (two horizontal, pointing forward at about 45°, and one longitudinally oriented, pointing downward, also at 45° eccentricity) were sufficient for autonomous steering and altitude control of an airplane in a simple confined environment.

Inspiration was also taken from flying insects with regard to the information processing stage. Although the extraction of OF itself was not inspired by the EMD model (Sect. 3.3.2) due to its known dependency on contrast and spatial frequency (Sect. 5.2.1), OF detection was at the core of the proposed control strategies. An efficient algorithm for OF detection was adapted to fit the embedded microcontrollers (Sect. 5.2). We showed that, as in flying insects, expanding optic flow could be used to sense proximity of objects and detect impending collisions. Moreover, ventral optic flow was a cue to perceive altitude above ground. The attractive feature of such simple solutions for depth perception is that they do not require explicit measurement of distance or time-to-contact, nor do they rely on accurate knowledge of the flight velocity. Furthermore, is has been shown that, in certain cases, they intrinsically adapt to the flight situation by triggering warnings farther away from obstacles that appear to be harder to avoid (Sect. 6.1.1). Another example of bio-inspired information processing is the fusion of gyroscopic information with vision. Although the simple scalar summation employed in our robots is probably far from what actually happens in the fly's nervous system, it is clear that some important interactions between visual input and halteres' feedback exist in the insect (Sect. 3.3.3).

At the behavioural level, the first steering strategy using a series of straight sequences interspersed with rapid turning actions was directly inspired by flies' behaviour (Sect. 3.4.3). The altitude control demonstrated on wheels relied on mechanisms inferred from experiments with honeybees. Such bees have been shown to regulate the experienced OF in a number of situations (Sect. 3.4.4). In the latest experiment, though, no direct connection with identified flies' behaviour can be advocated. Nonetheless, it is worth noticing that reflective control strategies, such as that proposed by

Braitenberg [1984], are likely to occur in many animals and although they have not yet been explicitly used by biologists to explain flight control in flying insects, they arguably constitute good candidates.

Finally, bio-inspiration was of great help in the design of our autonomous, vision-based flying robots. However, a great deal of engineering insight was required to tweak biological principles so that they could meet the final goal. It should also be noted that biology often lacks synthetic models, sometimes because biologists not having enough of an engineering attitude (see Wehner, 1987, for an interesting discussion), and sometimes due to an insufficiency of experimental data. For instance, biologists are just starting to study neuronal computation in flies with natural, behaviourally relevant stimuli [Lindemann *et al.*, 2003]. Such investigations will probably question many principles established so far with simplified stimuli [Egelhaaf and Kern, 2002]. Moreover, mechanical structures of flying robots as well as their processing hardware will never perfectly match biological systems. These considerations compelled us to explore an alternative approach to biomimetism, which takes inspiration from biology at the level of the Darwinian evolution of the species, as can be seen in the next Chapter.

CHAPTER 7

Evolved Control Strategies

In this Chapter things get slightly out of hand. You may regret this, but you will soon notice that it is a good idea to give chance a chance in the further creation of new brands of vehicles. This will make available a source of intelligence that is much more powerful than any engineering mind.

V. Braitenberg, 1984

This Chapter explores alternative strategies for vision-based navigation that meet the constraints of ultra-light flying robots: few computational resources, very simple sensors, and complex dynamics. A genetic algorithm is used to evolve artificial neural networks that map sensory signals into motor commands. A simple neural network model has been developed, which fits the limited processing power of our lightweight robots and ensures real-time capability. The same sensory modalities as in Chapter 6 were used, whereas information processing strategies and behaviours were automatically developed by means of artificial evolution. First tested on wheels with the *Khepera*, this approach resulted in a successful vision-based navigation, that did not rely on optic flow. Instead, the evolved controllers simply measured the image contrast rate to steer the robot. Building upon this result, neuromorphic controllers were then evolved for steering the *Blimp2b*, resulting in efficient trajectories maximising forward translation while avoiding contacts with walls and coping with stuck situations.

7.1   Method

7.1.1   Rationale

One of the major problems facing engineers willing to use bio-inspiration in the process of hand-crafting artificial systems is the overwhelming amount of details and varieties of biological models. In the previous Chapter, we selected and adapted the principles of flying insects that seemed the most relevant to our goal of designing autonomous robots. However, it is not obvious that the use of optic flow as a visual preprocessing is the only alternative for these robots to navigate successfully. The control strategies consisting of using saccades or proportional feedback are equally questionable. It may be that other strategies are better adapted to the available sensors, processing resources, and dynamics of the robots.

This Chapter is an attempt to keep open the question of how sensory information should be processed, as well as what the best control strategy is in order to fulfil the initial requirement of "maximising forward translation", without dividing it into a set of control mechanisms such as course stabilisation, collision avoidance, etc. To achieve this, we use the method of evolutionary robotics (ER). This method allows us to define a substrate for the control system (a *neural network*[1]) containing free parameters (*synaptic weights*) that must be adapted to satisfy a performance criterion (*fitness function*) while the robot moves in its environment. In our application, the interest of this method is threefold:

- It allows us to fit the embedded microcontroller limitations (no floating point, limited computational power) by designing adapted *artificial neurons* (computational units of a neural network) before using evolution to interconnect them.

- It allows us to specify the task of the robot ("maximising forward translation") by means of the fitness function while avoiding specifying the details of the strategies that should be used to accomplish this task.

---

[1] Although other types of control structures can be used, the majority of experiments in ER employ some kind of artificial neural networks since they offer a relatively smooth search space and are a biologically plausible metaphors of mechanisms that support animal behaviours [Nolfi and Floreano, 2000].

- It implicitly takes into account the sensory constraints and dynamics of the robots by measuring their fitness while they are actually moving in the environment.

The drawback of ER with respect to hand-crafting bio-inspired controllers is that it requires a large amount of evaluations of randomly initialised controllers. To cope with this issue, we first rely on the *Khepera* robot (see Sect. 4.1.1) that is able to support any type of random control and withstand shocks against walls. Moreover, it is externally powered, i.e. it does not rely on batteries. This wheeled platform allows us to test and compare various kinds of visual preprocessing and parameters of evolution. The next step consists in building upon the results obtained on wheels to tackle the more complex dynamics typical of flying robots. Since the airplanes cannot support random controllers as this would very probably lead to a crash, we use the *Blimp2b* (see Sect. 4.1.2) as an intermediate flying platform. This platform already features much more complex dynamics than the *Khepera* robot, while still being able to withstand repetitive collisions. Moreover, a complete dynamic model has been developed [Zufferey *et al.*, 2006], which enables accurate simulation and faster evolutionary experiments. Since obtaining good solutions in simulation is not a goal *per se*, evolved controllers are systematically tested on the real *Blimp2b* at the end of the evolutionary process.

In addition to maximising forward translation, these two platforms (*Khepera* and *Blimp2b*) enable us to consider a corollary aspect of basic navigation: how to get out of stuck situations. Flying systems such as blimps can indeed get stuck in a corner of the test arena and be unable to maintain their forward motion as requested by the fitness function. This could not be tackled in the previous Chapter since (i) the airplanes could not be positioned in such a situation without an immediate crash resulting and (ii) optic flow only provides information when the robot is in motion. The robots selected as testbeds in this Chapter are able to both stop and reverse their course. An interesting question is thus whether evolved controllers can manage that kind of critical situations and, if so, what visual cues they use. Note that there is no need for modifying the global performance criterion of "maximising forward translation" in order to tackle this issue. It is sufficient to start each evaluation period with the robot in such a critical situation. If the robot cannot quickly get out of it, it will not be able to move forward during the rest of the evaluation period, thus leading to a very low fitness.

7.1.2 Evolutionary Process

An initial *population* of different *individuals*, each represented by the *genetic string* that encodes the parameters of a neural controller, is randomly created. The individuals are evaluated one after the other on the same physical (or simulated) robot. In our experiments, the population is composed of 60 individuals. After ranking the individuals according to their performance (using the fitness function, see Sect. 7.1.4), each of the top 15 individuals produces 4 copies of its genetic string in order to create a new population of the same size. The individuals are then randomly paired for *crossover*. One-point crossover is applied to each pair with 10 % probability and each individual is then mutated by switching the value of a bit with a probability of 1 % per bit. Finally, a randomly selected individual is substituted by the original copy of the best individual from the previous generation (*elitism*). This procedure is referred to as a rank-based truncated selection, with one-point crossover, bit *mutation*, and elitism [Nolfi and Floreano, 2000].

Each individual of the population is evaluated on the robot for a certain amount T of sensory-motor cycles (each lasting from 50 to 100 ms). The length of the *evaluation period* is adapted to the size of the arena and the typical robot velocity, in order for the individuals to have a chance to experience a reasonable amount of situations. In practice, we use an evaluation period of 40 to 120 s (or 400 to 2400 sensory-motor cycles). Usually, at least two evaluations are carried out with the same individual in order to average the effect of different starting positions on the global fitness.

This evolutionary process is handled by the software *goevo* (Sect. 4.3.1) that manages the population of genetic strings, decodes each of them into an individual with its corresponding neural controller, evaluates the fitness and carries out the selective reproduction at the end of the evaluation of the whole population. Two operational modes are possible (Fig. 7.1). In the *remote mode*, the neural controller (called *PIC-NN* for PIC-compatible neural network) is emulated within *goevo*, which exchanges data with the robot every sensory-motor cycle. In the *embedded mode*, the neural controller is implemented within the microcontroller of the robot and data exchanges occur only at the beginning and at the end of the evaluation periods. The remote mode allows the monitoring of the internal state of the controller whereas the embedded mode ensures a full autonomy of the robot at the end of the evolutionary process.

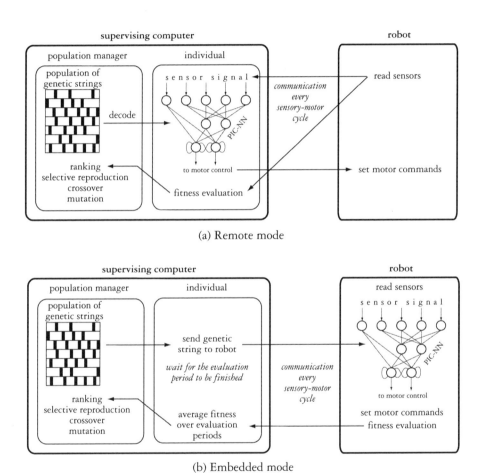

Figure 7.1 Two possible modes of operation during evolutionary runs. (a) Remote mode: the neural network (called PIC-NN) is run in the supervising computer that asks the robot for sensor values at the beginning of every sensory-motor cycle and sends back the motor commands to the robot. (b) Embedded mode: PIC-NN is embedded in the robot microcontroller and communication occurs only at the beginning and at the end of an evaluation period.

The advantage of the remote mode is that the monitoring of the network's internal state is straightforward and that it is easier to debug and modify the code. However, the need for sending all sensor values at every cycle is a weakness since this takes time (especially with vision) and thus lengthens the sensory-motor cycle. Furthermore, once the evolutionary process has ended, the best evolved controller cannot be tested without the supervising computer, i.e. the robot is not truly autonomous. In contrast, in the embedded mode, there is a lack of visibility with regard to the internal state of the controller. However, the sensory-motor cycle time can be reduced and once a genetic string is downloaded, the robot can work on its own for hours without any communication with an off-board computer.

In order to ensure the flexibility with respect to the type and the phase of experiment to be carried out, both modes are possible within our framework and can be used as required. It is also possible to carry out an evolutionary run in remote mode and to test good controllers in embedded mode only at the end. Furthermore, it is very useful to have the remote mode when working with a simulated robot that does not possess a microcontroller.

7.1.3   Neural Controller

An artificial *neural network* is a collection of units (*artificial neurons*) linked by weighted connections (*synapses*). Input units receive sensory signals and output units control the actuators. Neurons that are not directly connected to sensors or actuators are called *internal units*. In its simplest form, the output of an artificial neuron y_i (also called *activation value* of the neuron) is a function Λ of the sum of all incoming signals x_j weighted by *synaptic weights* w_{ij}:

$$y_i = \Lambda \left(\sum_{j}^{N} w_{ij} x_j \right), \tag{7.1}$$

where Λ is called the *activation function*. A convenient activation function is $\tanh(x)$ because for any sum of the input, the output remains within the range $[-1, +1]$. This function acts as a linear estimator in its center region (around zero) and as a threshold function in the periphery. By adding an incoming connection from a *bias unit* with a constant activation value of -1, it is possible to shift the linear zone of the activation function by modifying the synaptic weight from this bias.

In the targeted ultra-light robots, the neural network must fit the computational constraints of the embedded microcontroller. The PIC-NN (Fig. 7.2a) is thus implemented using only integer variables with limited range, instead of using high-precision floating point variables as it is usually the case when neural networks are emulated on desktop computers. Neuron activation values (outputs) are coded as 8-bit integers in the range $[-127, +127]$. The PIC-NN activation function is stored in a lookup table with 255 entries (Fig. 7.2c) so that the microcontroller does not have to compute the tanh function at every update. Synapses multiply activation values by an integer factor w_{ij} in the range $[-7, +7]$ which is then divided by 10 to ensure that a single input cannot saturate a neuron on its own. The range has been chosen to encode each synaptic weight on 4 bits (1 bit for the sign, 3 bits for the amplitude). Although activation values are 8-bit signed integers, the processing of the weighted sum (Fig. 7.2b) is done on a 16-bit signed integer to avoid overflows. The result is then limited to $[-127, +127]$ in order to get the activation function result through the look-up table.

The PIC-NN is a discrete-time, recurrent neural network, whose computation is executed once per sensory-motor cycle. Recurrent and lateral connections use the pre-synaptic activation values from the previous cycle as input. The number of input and internal units, the number of direct connections from input to output, and the activation of lateral and recurrent connections can be freely chosen. Since each synapse of a PIC-NN is encoded on 4 bits, the corresponding binary genetic string is thus composed of the juxtaposition of the 4-bit blocks, each represented by a gray square in the associated connectivity matrix (Fig. 7.2d).

In all experiments presented in this Chapter, the PIC-NN had 2 internal neurons and 2 output neurons whose activation values were directly used to control the actuators of the robot (positive values correspond to a positive rotation of the motor, whereas negative values yield a negative rotation). The two internal neurons were inserted in the hope that they could act as a stage of analysis of the incoming visual input in order to provide the output layer with more synthetic signals. Recurrent and lateral

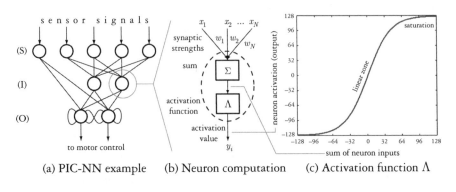

(a) PIC-NN example   (b) Neuron computation   (c) Activation function Λ

(d) PIC-NN connectivity matrix

Figure 7.2 The PIC-NN. (a) The architecture of the PIC-NN. Sensor input units are denoted S, and input and output neurons are labelled I and O, respectively. The bias unit B is not shown. In this example, recurrent and lateral connections are present among output neurons. One input unit is directly connected to the output units, whereas four other input neurons are connected to the internal units. (b) Details of the computation occurring in a single neuron. Note that only internal and output neurons have this computation. Input units have an activation value proportional to their input. (c) A discrete activation function implemented as a lookup table in the microcontroller. (d) The PIC-NN connectivity matrix. Each gray square represents one synaptic weight. Each line corresponds either to an internal or an output neuron. Every column corresponds to one possible pre-synaptic unit: either neurons themselves, or input units, or the bias unit. The lateral and recurrent connections (on the diagonal of the left part of the matrix) can be enabled on the internal and/or output layers. In this implementation, the output neurons never send their signal back to the internal or input layers. Input units can either be connected to the internal layer or directed to the output neurons.

connections were enabled only in the output layer thus permitting an inertia or low-pass filtering effect on the signals driving the motors. The number of input units depended on the type of sensory preprocessing.

7.1.4 Fitness Function

The design of a *fitness function* for the evaluation of the individuals is a central issue in any evolutionary experiment. In our experiments, we relied on a fitness function that is measurable by sensors available onboard the robots, as well as sufficiently simple to avoid unwanted pressure toward specific behaviours (e.g. sequences of straight movements and rapid turning actions). The fitness was simply a measure of forward translation.

For the *Khepera*, the instantaneous fitness was the average of the wheel speeds (based on wheel encoders):

$$\Phi_{Khepera}(t) = \begin{cases} \dfrac{v_L(t) + v_R(t)}{2} & \text{if } v_L(t) + v_R(t) > 0\,, \\ 0 & \text{otherwise}\,, \end{cases} \qquad (7.2)$$

where v_L and v_R are the left and right wheel speeds, respectively. They were normalised with respect to their maximum allowed rotation rate (corresponding to a forward motion of 12 cm/s). If the *Khepera* rotated on the spot (i.e. $v_L = -v_R$), the fitness was zero. If only one wheel was set to full forward velocity, while the other one remained blocked, the fitness reached 0.5. When the *Khepera* tried to push against a wall, its wheels were blocked by friction, resulting in null fitness since the wheel encoders would read zero.

In order to measure forward translation of the *Blimp2b*, we used the anemometer located below its gondola (Fig. 4.2). The instantaneous fitness can thus be expressed as:

$$\Phi_{Blimp}(t) = \begin{cases} v_A(t) & \text{if } v_A(t) > 0\,, \\ 0 & \text{otherwise}\,, \end{cases} \qquad (7.3)$$

where v_A is the output of the anemometer, which is proportional to the forward speed (the direction in which the camera is pointing). Moreover, v_A was normalised with respect to the maximum value obtained during straight motion at full speed. Particular care was taken to ensure that the anemometer was outside the flux of the thrusters to avoid it rotating for example when the blimp was pushing against a wall. Furthermore, no significant rotation of the anemometer was observed when the blimp rotated on the spot.

The instantaneous fitness values given in equations (7.2) and (7.3) were then averaged over the entire evaluation period:

$$\bar{\Phi} = \frac{1}{T} \sum_{t=1}^{T} \Phi(t), \qquad (7.4)$$

where T is the number of sensory-motor cycles of a trial period. For both robots, a fitness of 1.0 would thus correspond to a straight forward motion at maximum speed for the entire duration of the evaluation period. However, this cannot be achieved in our test environments (Figs 4.15 and 4.16) where the robots have to steer in order to avoid collisions.

7.2   Experiments on Wheels

We first applied the method to the *Khepera* to determine whether evolution could produce efficient behaviour when the PIC-NN was fed with raw vision. The results were then compared to the case when optic-flow is provided instead. We then tackled the problem of coping with stuck situations. These results on wheels constituted a good basis for further evolutionary experiments in the air with the *Blimp2b* (Sect. 7.3).

All the experiments in this Section were carried out on the *Khepera* equipped with the *kevopic* extension turret and the frontal 1D camera in the 60 × 60 cm textured arena (Fig. 4.15a). An evaluation period lasted 40 s (800 sensory-motor cycles of 50 ms) and was repeated two times per individual. The fitnesses of the two evaluation periods were then averaged. The resulting fitness graphs were based on an average of 3 evolutionary runs starting from a different random initialisation of the genetic strings.

7.2.1   Raw Vision versus Optic Flow

To answer the questions of whether optic flow and/or saccadic behaviour are required (see Sect. 7.1.1), two comparative experiments were set up. In the first one, called "raw vision", the entire image was fed to the neural controller without any temporal filtering[2], whereas in the second, called "optic flow", four optic-flow detectors (OFDs, see Sect. 5.2.5) served as exclusive visual input to the neural controller (Fig. 7.3). The initialisation procedure before each evaluation period consisted of a routine where the *Khepera* drove away from the walls for 5 s using its proximity sensors (see Sect. 4.1.1). We could thus avoid dealing with the corollary question of whether evolved individuals can manage stuck situations such as frontally facing a wall This is tackled in the next Section.

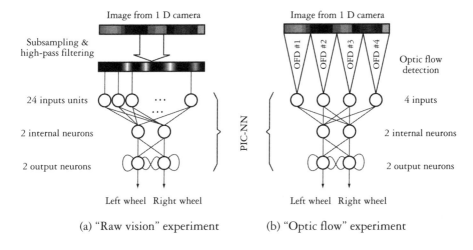

(a) "Raw vision" experiment   (b) "Optic flow" experiment

Figure 7.3   The configuration of visual preprocessing and PIC-NN for the comparison between "raw vision" and "optic flow". (a) 50 pixels from the center of the 1D camera are subsampled to 25 and high-pass filtered with a rectified spatial difference for every neighbouring pixel. The resulting 24 values are directly sent to the 24 inputs of the PIC-NN. (b) 48 pixels are divided into 4 regions of 12 pixels, on which the image interpolation algorithm (I2A, see Sect. 5.2.3) is applied. The optic-flow detector (OFD) outputs are then passed on to the 4 inputs of the underlying PIC-NN.

---

[2] As opposed to optic-flow processing, which involves a spatio-temporal filter (see equations 5.5 and 5.8).

The first experiment with "raw vision" capitalised on existing results and was directly inspired by the experiment reported by Floreano and Mattiussi [2001], where a *Khepera* was evolved for vision-based navigation in the same kind of textured arena. The main difference between this experiment and the one presented herein concerns the type of neural network.[3] The controller used by Floreano and Mattiussi [2001] was a spiking neural network emulated in an off-board computer (remote mode)[4] instead of a PIC-NN. The idea of high-pass filtering vision before passing it on to the neural network has been maintained in this experiment, although the processing was carried out slightly differently in order to reduce computational costs.[5] The main reason for high-pass filtering the visual input was to reduce dependency on background light intensity.[6]

In the second experiment with optic-flow, the parameters remained unchanged, except the visual preprocessing and the number of input units in the PIC-NN. Note that the two external OFDs had exactly the same configuration as in the optic-flow based steering experiment (Fig. 6.8). Therefore, this visual information together with a saccadic behaviour should, in principle, be enough to efficiently steer the robot in the test arena.

Results

The graph in Figure 7.4(a) shows the population's mean and best fitness over 30 generations for the case of "raw vision". The fitness rapidly improved in the first 5 generations and then gradually reached a plateau of about 0.8 around the 15th generation. This indicates that evolved controllers

---

[3] Other minor differences concern the vision module (see Zufferey *et al.*, 2003), the number of used pixels (16 instead of 24), the details of the fitness function, and the size of the arena.

[4] More recent experiments have demonstrated the use of simpler spiking networks for embedded computation in a non-visual task [Floreano *et al.*, 2002]. See Floreano *et al.* [2003] for a review.

[5] Instead of implementing a Laplacian filter with a kernel of 3 pixels $[-.5 \ 1 \ -.5]$, we here used a rectified spatial difference of each pair of neighbouring pixels, i.e. $|I(n) - I(n-1)|$, where n is the pixel index and I the intensity. The outcome was essentially the same, since both filters provide a measure of local image gradient.

[6] Although the test arenas were artificially lit, they were not totally protected from natural light from outdoors. The background light intensity could thus fluctuate depending on the position of the sun and the weather.

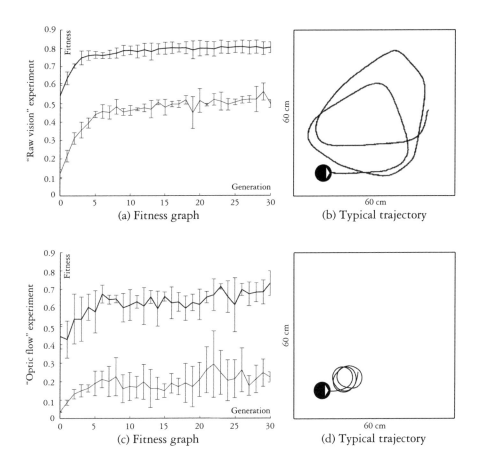

Figure 7.4 Comparative results from the "raw vision" and "optic flow" experiments. (a) & (c) Mean (thin line) and best (thick line) population fitness for 30 generations. The data points are averages over three evolutionary runs and the error bars are the standard deviations among these three runs. (b) & (d) A typical trajectory of the best individual is plotted based on data from the wheel encoders.

found a way of moving forward while avoiding to get stuck against the surrounding walls. However, the fitness value does not inform us about the specific behaviour adopted by the robot. To obtain this information, the best evolved controller was tested and its wheel encoders recorded in order to reconstruct the trajectory. Figure 7.4(b) shows that the robot moved along a looping trajectory, whose curvature depends on the visual input.[7]

---

[7] The resulting behaviour is very similar to that obtained by Floreano and Mattiussi [2001].

Note that this behaviour is not symmetrical. Evolution found a strategy consisting in always turning in the same direction (note that the initial direction can vary between experiments) and adapting the curvature radius to exploit the available space of the arena. In this experiment, the best evolved controllers always set their right wheel to full speed, and controlled only the left one to steer the robot. This strategy is in contrast with the hand-crafted solution implemented in Section 6.1.3, which consisted in going straight and avoiding walls at the last moment and in the direction opposite to the closest side.

With "optic flow" as visual input, the resulting fitness graph (Fig. 7.4c) displays significantly lower maximum values as compared to the previous experiments. The resulting trajectory (Fig. 7.4d) reveals that only a very minimalist solution, where the robot rotates in small circles, is found. This is not even vision-based navigation, since visual input does not have any influences on the constant turning radius. This strategy can, however, still produce a relatively high fitness of almost 0.7 since the individuals were always initialised far from the walls at the beginning of the evaluation periods and thus had enough space for such movement independently of their initial heading.

Discussion

Evolution with optic-flow as visual preprocessing did not produce acceptable navigation strategies, despite that the neural controller was provided with the same kind of visual input as that described in Section 6.1.3. This can be explained by the fact that OFDs give useful information only when the robot is moving in a particular manner (straight forward at almost constant speed), but since the output of the neural networks used here depended solely on the visual input, it is likely that a different neural architecture would be needed to properly exploit information from optical flow. It should be noted that we did not provide derotated OF to the neural network in this experiment. We hoped that the evolved controller could find a way of integrating the rotational velocity information based on the left and right wheel speeds $(v_L - v_R)$, which are produced by the neural network itself. However, this did not happen.

In contrast, evolution with "raw vision" produced interesting results with this simple PIC-NN. In order to understand how the visual information could be used by the neural network to produce the efficient behaviour, we made the hypothesis that the controller relied essentially on the contrast rate present in the image (a spatial sum of the high-pass filtered image). To test this hypothesis, we plotted the rotation rate $(v_L - v_R)$ as a function of the spatial average of the visual input (after high-pass filtering) over the entire field of view (FOV) while the individual was moving freely in the arena.

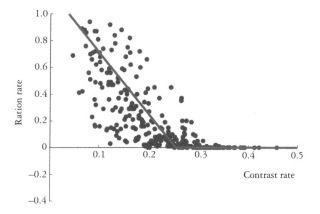

Figure 7.5   The *Khepera* rotation rate versus the image contrast rate during normal operation of the best evolved individual in its test arena. The contrast rate is the spatial average of the high-pass filter output (a value of 1.0 would correspond to an image composed exclusively of alternately black and white pixels). The rotation rate is given by $(v_L - v_R)$, where v_L and v_R are normalised in the range $[-1, +1]$.

The resulting graph (Fig. 7.5) shows that an almost linear relation existed between the contrast rate over the entire image and the rotation rate of the *Khepera*. In other words, the robot tended to move straight when a lot of contrast was present in the image, whereas it increased its turning rate as soon as less contrast was detected. The dispersion of the points in the right part of the graph shows that the processing of this particular neural network cannot be exclusively explained by this strategy. In particular, it is likely that some parts of the image are given more importance than others in the steering process. However, this simple analysis reveals the underlying logic of the evolved strategy, which can be summarised as follows: "move straight when the contrast rate is high, and increase the turning rate linearly with a decreasing contrast rate" (see the thick gray lines in Figure 7.5).

In summary, rather than to rely on optic flow and symmetrical saccadic collision avoidance, the successful controllers employed a purely spatial property of the image (the contrast rate) and produced smooth trajectories to circumnavigate the arena in a single direction.

7.2.2   Coping with Stuck Situations

This Section tackles the critical situations that occur when the robot faces a wall (or a corner). The issue was handled by adopting a set of additional precautions during the evolutionary process. Concurrently, we built upon the previous results in order to decrease the number of sensory input to the PIC-NN. This decreased the size of the genetic string and accelerated the evolutionary process.

Additional Precautions

In order to force individuals to cope with critical situations without fundamentally changing the fitness function, a set of three additional precautions were taken:

- Instead of driving the robot away from walls, the initialisation procedure placed them against a wall by driving them straight forward until one of the front proximity sensors became active.
- The evaluation period was prematurely interrupted (after 5 s) if the individual did not reach at least 10 % of the maximum fitness (i.e. 0.1).
- The instantaneous fitness function $\Phi(t)$ was set to zero whenever a proximity sensor (with a limited range of about 1-2 cm) became active.

Visual Preprocessing

Since the evolution of individuals with access to the entire image has mainly relied on the global contrast rate in the image (see discussion of Section 7.2.1), we then deliberately divided the image into 4 evenly distributed regions and computed the contrast rate in each of them before feeding the neural controller with the resulting values (Fig. 7.6). We call this kind of pre-processing associated with the corresponding image region *contrast rate de-*

tector (CRD). Since the high-pass spatial filtering is a kind of edge enhancement, the output of such a CRD is essentially proportional to the number of edges seen in the image region. This preprocessing reduced the size of the neural network with respect to the "raw vision" approach and thus limited the search space of the genetic algorithm[8]. Since the additional precautions already rendered the task more complex, the reduction of the search space was not expected to yield significant acceleration in the evolutionary process. However, this would help maintain the number of required generations at a reasonable amount.

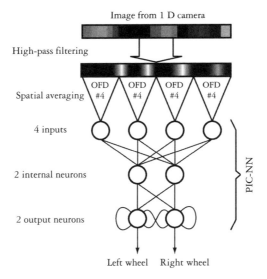

Figure 7.6 Visual preprocessing and PIC-NN for the experiment with critical starting situations. The intensity values from the 1D camera are first high-pass filtered with a rectified spatial difference for every other neighbouring pixels. The spatial averaging over 4 evenly distributed regions occurs in order to feed the 4 input units of the PIC-NN.

Results

The resulting fitness graph (Fig. 7.7a) is similar to the one from the "raw vision" experiment (Fig. 7.4a). Although progressing slightly slower in the first generations, the final maximum fitness values after 30 generations were

---

[8] The genetic string encoding the PIC-NN measured 80 bits instead of 240 bits in the "raw vision" experiment.

identical, i.e. 0.8. The increased difficulty of the task due to the additional precautions is indicated in the fitness graph by the lower average fitness over the population (approx. 0.35 instead of 0.5).

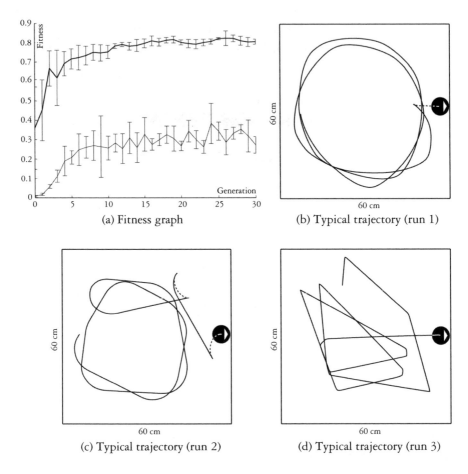

(a) Fitness graph

(b) Typical trajectory (run 1)

(c) Typical trajectory (run 2)

(d) Typical trajectory (run 3)

Figure 7.7 Results of the evolutionary experiment with the *Khepera* using 4 contrast rate detectors and coping with critical starting situations. (a) Mean (thin line) and best (thick line) population fitness for 30 generations. The data points are averages over three evolutionary runs. (b)-(d) Typical trajectories of the best individuals of the 3 runs. The *Khepera* (black circle with the white arrow indicating the forward direction) is always placed perpendicularly facing a wall at the beginning of the experiment to demonstrate its ability to rapidly get out of this difficult situation. A dotted trajectory line indicates backward motion.

The genetic algorithm found a way of coping with the new set of precautions in spite of the limited number of sensory inputs. In order to better demonstrate the higher robustness obtained in this experiment, the typical trajectories of the best evolved individuals of various evolutionary runs were plotted with the *Khepera* starting against a wall (and facing it). We observed a number of different behaviours that produced the same average fitness values. In all cases, the individuals managed to quickly escape from the critical starting position, either by backing away from the wall (Fig. 7.7b-c) during a short period of time (roughly 2 s) or rotating on the spot until finding a clear path (Fig. 7.7d). Once escaped, they quickly recovered a forward motion corresponding to high fitness. The behaviours either consisted in navigating in large circles and slightly adapting the turning rate when necessary (Fig. 7.7b), or moving in straight segments and steering only when close to a wall. In this latter case, the individuals either described smooth turns (Fig. 7.7c) or on-the-spot rotations (Fig. 7.7d). The individuals that rotated on the spot when facing a wall sometimes exploited the same strategy in order to avoid collisions later on.

These results demonstrated that a range of possible strategies was possible, and that they all fulfilled the basic requirement of "maximising forward translation" even if the starting position was critical (i.e. requires a specific behaviour that is not always used later on). Rather than using optic-flow, these strategies relied on spatial properties (contrast rate) of the visual input.

7.3   Experiments in the Air

A preliminary set of experiments carried out solely on a physical blimp [Zufferey *et al.*, 2002] indicated that artificial evolution could generate, in about 20 generations, neuromorphic controllers able to drive the flying robot around the textured arena. However, the obtained strategies largely relied on contacts with walls to stabilise the course of the blimp in order to gain forward speed. It should be noted that the *Blimp1* (ancestor of the current *Blimp2b*) used in this preliminary experiments was significantly less manoeuvrable and had no rate gyro. Later on, the *Blimp2* (very similar to the

Blimp2b) equipped with a yaw rate gyro (whose output was passed on to the neural controller) produced smoother trajectories without using the walls for stabilisation [Floreano *et al.*, 2005]. These evolutionary runs performed directly on the physical flying robots were rather time-consuming. Only 4 to 5 generations could be tested in one day (the battery had to be changed every 2-3 hours) and more than one week was required to obtain successful controllers. Additionally, certain runs had to be dismissed because of mechanical problems such as motor deficiencies.

After these early preliminary experiments, the simulator was developed in order to accelerate and facilitate the evolutionary runs (see Sect. 4.3.2). In contrast to previous experiments with *Blimp1* and *Blimp2*, we here present experiments with the *Blimp2b* where

- the evolution took place entirely in simulation and only the best evolved controllers were transferred to the real robot,
- the same set of precautions as those developed with the *Khepera* (see Sect. 7.2.2) were used to force individuals to cope with critical situations (facing a wall or a corner),
- a set of virtual[9] proximity sensors were used during simulated evolution to set the instantaneous fitness to zero whenever the blimp was close to a wall (part of the above-mentioned precautions).

This Section is divided into two parts. First the results obtained in simulation are presented, and then the transfer to reality of the best evolved individual is described.

7.3.1 Evolution in Simulation

The neural controller was evolved in order to steer the *Blimp2b* in the square arena (Fig. 4.16) by use of only visual and gyroscopic information available from on-board sensors.[10] As for the latest experiment with the *Khepera* (see Sect. 7.2.1), the visual input was preprocessed with 4 CRDs, which fed the

---

[9] A virtual sensor is a sensor implemented only in simulation, and that does not exist on the real robot.

[10] In these experiments, the altitude was not under evolutionary control, but was automatically regulated using information from the distance sensor pointing downward (see Sect. 4.1.2).

PIC-NN (Fig. 7.8). In addition, the pixel intensities coming from the 1D camera were binarised. Since the visual surrounding, both in simulation and reality, was black and white, thresholding the image ensured a better match among the two worlds.

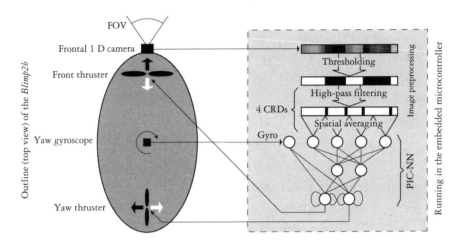

Figure 7.8 Left: An outline of the sensory inputs and actuators of the *Blimp2b*. Right: The neural network architecture and vision preprocessing.

Since one of the big differences between the *Khepera* and the *Blimp2b* was the need for course stabilisation (see Table 4.3), the yaw rate gyro output was also provided to the neural controller. This additional sensory information was sent to the PIC-NN via an input unit, which was directly connected to the output neurons. The motivation for this direct connection was that, based on results obtained in Section 6.1.4, a simple proportional feedback loop connecting the rate gyro to the rudder of the airplane was sufficient to provide course stabilisation.

The PIC-NN thus had 4 visual input units connected to the internal layer, 1 gyro input unit directly connected to the output layer, 2 internal neurons, and 2 output neurons controlling the frontal and yaw thrusters (Fig. 7.8). The PIC-NN was updated every sensory-motor cycle lasting

100 ms in reality.[11] The evaluation periods lasted 1200 sensory-motor cycles (or 2 min real-time).

As in the last experiment with the *Khepera* robot (see Sect. 7.2.2), a set of additional precautions were taken during the evolutionary process in order to evolve controllers capable of moving away from walls. The 8 virtual proximity sensors (Fig. 4.14) were used to set the instantaneous fitness to zero whenever the *Blimp2b* was less than 25 cm from a wall. In addition, individuals that displayed poor behaviours (less than 0.1 fitness value) were prematurely interrupted after 100 cycles (i.e. 10 s).

Results

Five evolutionary runs were performed, each starting with a different random initialisation. The fitness graph (Fig. 7.9a) displays a steady increase up to the 40th generation. Note that it was far more difficult for the *Blimp2b* to approach a fitness of 1.0 as opposed to the *Khepera* because of inertial and drag effects. However, all five runs produced efficient behaviours in less than 50 generations rendering it possible to navigate in the room in the forward direction while actively avoiding walls. Figure 7.9(b) illustrates the typical preferred behaviour of the best evolved individuals. The circular trajectory was, from a velocity point of view, almost optimal because fitting the available space well (the back of the blimp sometimes got very close to a wall without touching it). Evolved robots did not turn sharply to avoid the walls, probably because this would cause a tremendous loss of forward velocity. The fact that the trajectory was not centered in the room is probably due to the spatial frequency discrepancy among walls (two walls contained fewer vertical stripes than the other two). The non-zero angle between the heading direction of the blimp (indicated by the small segments) and the trajectory suggests that the simulated flying robot kept side-slipping and thus that the evolved controllers required to take into account the quite complex dynamics of the blimp by partly relying on air drag to compensate for the centrifugal force.

---

[11] A longer sensory-motor cycle than with the *Khepera* was chosen here, primarily because the communication through the radio system added certain delays. In embedded mode (without monitoring of parameters), the sensory-motor cycle could easily be ten times faster.

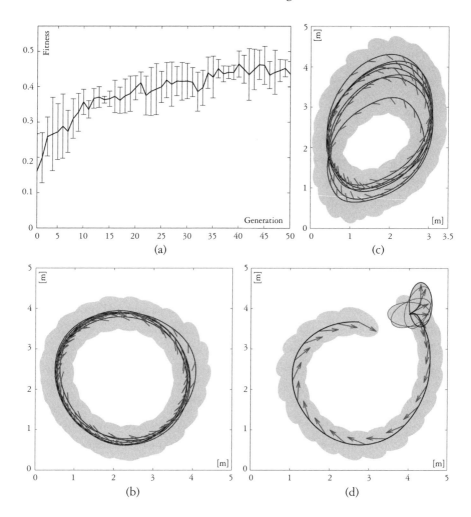

Figure 7.9 Results in simulation. (a) Average fitness values and standard deviations (over a set of five evolutionary runs) of the fittest individuals of each generation. (b) A top view of the typical trajectory during 1200 sensory-motor cycles of the fittest evolved individual. The black continuous line is the trajectory plotted with a time resolution of 100 ms. The small segments indicate the heading direction every second. The light-gray ellipses represent the envelope of the blimp also plotted every second. (c) The trajectory of the fittest individual when tested for 1200 sensory-motor cycles in a room that was artificially shrunk by 1.5 m. (d) When the best individual was started against a wall, it first reversed its front thruster while quickly rotating clockwise before resuming its preferred behaviour. The ellipse with the bold black line indicates the starting position, and the following ones with black outlines indicate the blimp envelope when the robot is in backward motion. The arrows indicate the longitudinal orientation of the blimp, irrespective of forward or backward movement.

In order to further assess the collision avoidance capability of the evolved robots, we artificially reduced the size of the room (another useful feature of the simulation) and tested the same individual (best performer) in this new environment. The blimp modified its trajectory into a more elliptic one (Fig. 7.9c), moving closer to the walls, but without touching them. In another test, where the best individual was deliberately put against a wall (Fig. 7.9d), it reversed its front thruster, and backed away from the wall while rotating in order to recover its preferred circular trajectory. This behaviour typically resulted from the pressure exerted during evolution by the fact that individuals could be interrupted prematurely if they displayed no fitness gain during the first 10 s. They were therefore constrained to develop an efficient strategy to get out from whatever initial position they were in (even at the expense of a backward movement, which obviously brought no fitness value) in order to quickly resume the preferred forward trajectory and gain fitness.

7.3.2   Transfer to Reality

When the best evolved neuromorphic controller was tested on the physical robot (without further evolution), it displayed an almost identical behaviour.[12] Although we were unable to measure the exact trajectory of the blimp in reality, the behaviour displayed by the robot in the 5×5 m arena was qualitatively very similar to the simulated one. The *Blimp2b* was able to quickly drive itself on its preferred circular trajectory, while robustly avoiding contact with the walls.

The fitness function could be used as an estimate of the quality of this transfer to reality. A series of comparative tests were performed with the best evolved controller, in simulation and reality. For these tests, the virtual proximity sensors were not used since they did not exist in reality. As a result, the instantaneous fitness was not set to zero when the blimp was close to a wall, as was the case during evolution in simulation. The fitness values were therefore expected to be slightly higher than those shown in the fitness graph of Figure 7.9(a). The best evolved controller was tested 10

---

[12] Video clips of simulated and physical robots under control of this specific evolved neural controller are available for download from http://book.zuff.info

times in simulation and 10 times in reality for 1200 sensory-motor cycles. The results from these tests, which are plotted in Figure 7.10, show that the controllers having evolved in simulation obtained very similar performances when assessed on the real testbed.

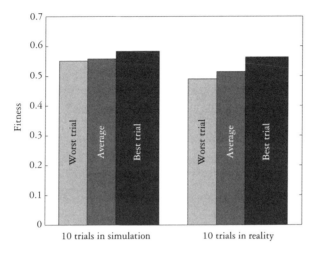

Figure 7.10 The performance when going from simulation to reality with the best controller. Fitness results from 10 trials with the best evolved individual; simulation to the left, reality to the right.

In order to further verify the correspondence between the simulated and real robot, we compared signals from the anemometer, the rate gyro and the actuators while the *Blimp2b* moved away from a wall. These signals provided an internal view of the behaviour displayed by the robot. The *Blimp2b* was thus started facing a wall, as shown in Figure 7.9(d), both in simulation and in reality. Figure 7.11 shows the very close match between signals gathered in reality and those recorded in an equivalent simulated situation. At the beginning, the front thruster was almost fully reversed while a strong yaw torque was produced by the yaw thruster. These actions yielded the same increment in rotation rate (detected by the rate gyro) and a slight backward velocity (indicated by negative values of the anemometer), both in reality and in simulation. After approximately 3 s, the blimp had almost finished the back-and-rotation manoeuvre and started a strong counter-action with the yaw thruster to cancel the yawing movement, thus resulting in a noticeable decrease in the rate gyro signal. Subsequently, the

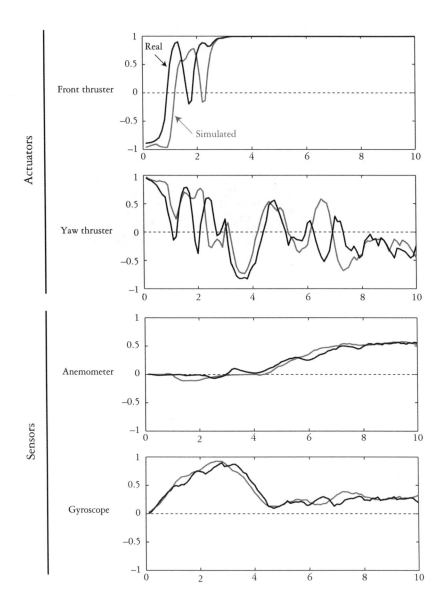

Figure 7.11 A comparison of thruster commands and sensor values between simulation and reality when the best evolved individual started in a position facing a wall, as shown in Figure 7.9(d). The thruster values are normalised with respect to the full range; the anemometer output is normalised with respect to the maximum forward velocity; the rate gyro data is normalised with respect to the maximum rotation velocity. Note that, already after 4 s, the robot started to accumulate fitness since the anemometer measured forward motion (during evolution, 10 s were allowed before interruption due to poor fitness).

robot accelerated forward (as shown in the anemometer graph) to recover its preferred circular trajectory (as revealed by the almost constant, though not null, rate gyro values). Slight discrepancies among signals from simulation and reality can be explained by variations in the starting position implying slightly different visual inputs, inaccuracies in sensor modelling, and omitted higher order components in the dynamic model [Zufferey *et al.*, 2006].

7.4 Conclusion

The present chapter took an interest in exploring alternative strategies to vision-based steering. An evolutionary robotics (ER) approach was chosen for its capability of implicitly taking care of the constraints related to the robot (sensors, processing power, dynamics) without imposing a specific manner of processing sensory information, nor forcing a pre-defined behaviour for accomplishing the task (maximising forward translation).

Artificial evolution was used to develop a neural controller mapping visual input to actuator commands. In the case of the *Khepera* robot, evolved individuals displayed efficient strategies for navigating the square textured arenas without relying on optic flow. The strategies employed visual contrast rate, which is a purely spatial property of the image. When the same neural controller was explicitly fed with optic flow, evolution did not manage to develop efficient strategies, probably as a result of optic flow requiring more delicate coordination between motion and perception than what could potentially be achieved with the simple neural network that was employed. Nevertheless, this result does not mean that there is no hope of evolving neural networks for optic-flow-based navigation. For instance, providing derotated optic flow and directing OFDs at equal eccentricity may prove beneficiary (see also Section 6.3).

When applied to the *Blimp2b*, artificial evolution found an efficient way of stabilising the course and steering the robot in order to avoid collisions. In addition, evolved individuals were capable of recovering from critical situations where they were incapable of simply moving forward to get a high fitness score.

These results were obtained using a neural network that was specifically developed in order to fit the low processing power of the embedded microcontroller, while ensuring real-time operation. The evolved controllers could thus operate without the help of any external computer. A ground station was required only during the evolutionary process in order to manage the population of genetic strings.

Comparison with Hand-crafting of Bio-inspired Control Systems

When using ER, the role of the designer is limited to the realisation of the robot, the implementation of the controller building blocks (in our case, artificial neurons), and the design of a fitness function. The evolutionary process then attempts to find the controller configuration that best satisfies all these constraints. The resulting strategies are interesting to analyse. In our case, we learnt that image contrast rate was a usable visual cue to efficiently drive our robots in their test arenas.

However, it is in some sense a minimalist solution that will work only under conditions equivalent to those existing during evolution. In particular, the individuals will fail as soon as the average spatial frequency of the surrounding texture changes. In contrast, the optic-flow-based control strategies developed in Chapter 6 were designed to be largely insensitive to spacial frequency. Also, the evolved asymmetrical behaviour should perform less efficiently in an elongated environment (e.g. a corridor), whereas the symmetrical collision avoidance strategy developed for the airplanes is better adapted to such a situation. To tackle these issues, it would be possible to change the environmental properties during evolution. This would however require longer evolutionary runs and probably more complex neural networks.

A significant drawback of ER with respect to hand-crafting bio-inspired controllers is that it requires a large amount of evaluations of randomly initialised controllers. To cope with this issue, the robot must be capable of supporting such controllers and recovering at the end of every evaluation period. If not, the use of an accurate, physics-based simulator is inevitable. The development of such a simulator can, depending on the dynamics of the robot, the complexity of the environment, and the type of sensors used, be quite difficult (see Nolfi and Floreano, 2000 for a detailed discussion about the use of simulation in ER).

Evolutionary Approach and Fixed-wing Aircraft

Airplanes such as the *F2* or the *MC2* would not support an evolutionary run for three reasons. First, they are not robust enough to withstand repeated collisions with the walls of the arena. Second, they cannot be automatically initialised into a good airborne posture at the beginning of each evaluation period. Third, they have a very limited endurance (approximately 10-30 min). The only solution for applying the evolutionary approach to such airplanes is to develop an accurate flight simulator. However, this is more difficult than with an airship, because, under the control of a randomly initialised neural controller, an airplane will not only fly in its standard regime (near level flight at reasonable speed), but also in stall situations, or high pitch and roll angles. Such non-standard flight regimes are difficult to model since unsteady-state aerodynamics play a predominant role.

To cope with this issue, certain precautions can be envisaged. For instance, it is conceivable to initialise the robot in level flight close to its nominal velocity and prematurely interrupt the evaluation whenever certain parameters (such as pitch and roll angles, and velocity) exceed a predefined range where the simulation is known to be accurate. This will also force the individuals to fly the plane in a reasonable regime.

Problems related to simulation-reality discrepancies could be approached with other techniques. Incremental evolution consisting of pursuing evolution in reality for a short amount of generations (see Harvey *et al.*, 1994 or Nolfi and Floreano, 2000, Sect. 4.4) could be a first solution, although a safety pilot would probably be required to initialise the aircraft and rescue it whenever the controller fails. Moreover, the procedure could be very time-consuming and risky for the robot. The second approach consists in using some sort of synaptic plasticity in the neural controller. Exploitation of synaptic adaptation has been shown to support fast self-adaptation to changing environments [Urzelai and Floreano, 2001].

Outlook

The present book describes the exclusive use of artificial evolution to set the synaptic strength of a simple neural network. However, artificial evolution in simulation could be employed to explore architectural issues such as airframe shape (provided that the simulator is able to infer the effects on the dynamics) or sensor morphology [Cliff and Miller, 1996; Huber *et al.*, 1996; Lichtensteiger and Eggenberger, 1999]. For instance, position and orientation of simple vision sensors could be left to evolutionary control and the fitness function could put some pressure toward the use of a minimum number of sensors. Ultimately, artificial evolution could also allow exploration of higher order combinations of behaviours (taking-off, flying, avoiding obstacles, going through small apertures, looking for food, escaping predators, landing, etc.). This research endeavour may even lead to an interesting comparison with existing models of how such behaviours are generated in insects.

CHAPTER 8

Concluding Remarks

I see insect level behavior as a noble goal for artificial intelligence practitioners. I believe it is closer to the ultimate right track than are the higher level goals now being pursued.

R. A. Brooks, 1986

The science fiction of an artificial flying insect buzzing around your office, suddenly deciding to escape through the door and managing to reach your colleague's room is a futuristic scenario that this book humbly contributed to bring closer to reality. The 10-gram microflyer demonstrated autonomous operation using only visual, gyroscopic and anemometric sensors. The signal processing and control were carried out entirely on-board, despite the plane's very limited payload of approximately 4 g. This has been made possible by developing ultra-light optic-flow detectors and fitting the algorithms (optic-flow detection and airplane control) in a tiny 8-bit microcontroller. The whole software running on the airplane used only a few thousand bytes of program memory. In flight, the airplane consumed less than 2 W, which is 30 times less than a desk light. Along a slightly different line, the blimp permitted the use of an evolutionary technique to automatically develop embedded neuromorphic controllers. This buoyant robot required only 1 W to autonomously circumnavigate the test arena while avoiding collisions with walls.

One of the main outcomes of this experimental exploration is the insight gained about the linking of simple visual features (such as local optic-flow or contrast rate) to actuator commands, in order to obtain efficient behaviour with lightweight and dynamic robots featuring limited computational resources. Typical problems arising when optic flow is used in absence of contact with an inertial frame (no odometry, unstable motion) have been solved by merging gyroscopic information and visual input. The results of the evolutionary experiments showed that optic flow is not the only way of processing monocular visual information for course control and collision avoidance. Although the evolved contrast-based solution cannot be generalised to other environments as easily as an optic-flow-based strategy, it represents an interesting alternative, requiring even less computational power.

Although the primary purpose of this research was to synthesize lightweight autonomous flyers, the size, energy, and computational constraints of the robotic platforms encouraged us to look at mechanisms and principles of flight control exploited by insects. Our approach to developing autonomous vision-based flying robots was inspired by biology at different levels: low-resolution insect-like vision, information processing, behaviour, neuromorphic controllers, and artificial evolution. In doing so, we have, in some sense, contributed to the testing of various biological models, in particular with the demonstration that an artificial flying insect could steer autonomously in a confined environment over a relatively long period of time. In this regard, this book is an illustration of the synergistic relationship that can exist between robotics and biology.

8.1   What's next?

The natural next step with indoor microflyers consists in getting out of the empty test arena and tackling unconstrained indoor environments such as offices and corridors with sparse furniture. In order to reach this objective, more attention will be required regarding the visual sensors. Two main challenges have to be addressed. First, the light intensity variations will be huge among different regions of such environments, some receiving direct sun light and others only artificially lit. Second, it may happen that parts of the visual surroundings have absolutely no contrast (e.g. white walls or windows). If the vision system samples the field of view only in very few regions (like it is currently the case in the robots presented in this book), it may provide no usable signals.

To cope with these issues, while maintaining the overall weight and power consumption as low as possible, is a challenge that may be tackled using custom-designed AVLSI[1] vision chips [Liu *et al.*, 2003] instead of using classical CMOS cameras. This technology provides a circuit-design approach to implementing certain natural computations more efficiently than standard logic circuits. The resulting chips usually consume at least 10 times less power than an equivalent implementation with CMOS imagers and digital processors. More specifically appealing for tackling the problem of background light fluctuations is the existence of adaptive photorecep-tor circuits [Delbrück and Mead, 1995] that automatically adapt to back-ground light over a very broad intensity range (more than 6 decades). These photoreceptors can be used as front-end to optic-flow detector circuits fitted on the same chip (e.g. Kramer *et al.*, 1995; Moeckel and Liu, 2007). This technology also provides the potential for widening the field of view in ar-ranging pixels and optic-flow circuits as desired on a single chip, while con-suming less energy and computational power than if the same functionality had to be achieved with standard CMOS sensors and vision processing in a microcontroller. As an example, Harrison [2003] presented an AVLSI chip for imminent collision detection based on the STIM model (see Sect. 3.3.3).

All in all, increasing the field of view and the number of optic flow de-tectors is definitely a research direction that is worth trying. However, even if custom-designed vision chips would allow specific arrangement of pixels, the issue of finding a lightweight optics providing a wide FOV remains. One solution may be to copy insects and develop artificial compound eyes. For instance, Duparre *et al.* [2004] have already developed micro-lens arrays that mimic insect vision. However, widening the FOV while working with flat image sensors is still an open issue with this technology.

Another approach would be to reduce the weight constraints by using outdoor flying platforms. In that case, standard vision systems with fish-eye lenses could be used to provide a 180° FOV in the flight direction. Image processing could be carried out on a far more powerful embedded processor

---

[1] Analog Very Large Scale Integration.

than the one used indoors. However, the payload for the required electronics would then increase to 50 g or so. Outdoor spaces, even urban canyons, allow for faster flight and thus increased payload, while remaining on the safe side with respect to people or buildings.

8.2   Potential Applications

Autonomous operation of ultra-light flying robots in confined environments without GPS nor active range finders is not trivial. This book explored a novel approach that yielded very dynamic behaviour, quite far from stabilised level flight between predefined way-points in open space, as it is often the case with current UAVs. The proposed biologically inspired solutions (from sensor suite to control strategies) are so lightweight and computationally inexpensive that they very much fit the growing demand for automating small UAVs that can fly at low altitude in urban or natural environments, where buildings, trees, hill, etc. may be present and a fast collision avoidance system is required [Mueller, 2001].

The described approach could also be of great help in automating even smaller flying devices, such as those presented in Section 2.1, which feature the same kind of properties as our flying robots: complex dynamics due to the absence of contact with an inertial frame, limited payload and restricted computational power. Distance sensors are not a viable solution in such cases, and visual and gyroscopic sensors are probably the best alternative to provide such robots with basic navigational skills. This approach seems to be acknowledged by the micromechanical flying insect (MFI) team in Berkeley, which is working on a sensor suite for their 25-mm flapping robot (Fig. 2.5). Although collision avoidance has not yet been tackled, preliminary work toward attitude control relies on visual (ocelli-like) and gyroscopic (halteres-like) sensors [Wu et al., 2003; Schenato et al., 2004].

More generally, the approach proposed in this book can provide low-level navigation strategies for all kinds of mobile robots charactesized by small size and non-trivial dynamics. It can equally be useful in a number of situations where the environment is unknown (no precise maps are available) and the use of GPS is not possible. This is the case, for example in indoor environments, underwater, or in planetary exploration, especially if the robot has to move close to the relief or in cluttered environments that are difficult to reconstruct with range finders.

Beyond the application of vision-based navigation strategies to mobile robots, the use of small indoor flying systems as tools for biological research can be envisaged. It is indeed striking that a growing number of biologists are assessing their models using physical robots (see, for instance, Srinivasan *et al.*, 1997; Duchon *et al.*, 1998; Lambrinos *et al.*, 2000; Rind, 2002; Reiser and Dickinson, 2003; Franceschini, 2004; Webb *et al.*, 2004). Until now, the used robotic platforms were terrestrial vehicles (e.g. Srinivasan *et al.*, 1998) or tethered systems (e.g. Reiser and Dickinson, 2003; Ruffier, 2004). An indoor flying platform with visual sensors and the ability to fly at velocities close to those reached by flies (1-3 m/s) potentially provides a more realistic testbed, one that can be used to assess models of visually guided behaviours in free flight. The fact that the aerial robots presented in this book fly indoors and are small and reasonably resistant to crashes would further ease the testing phase and alleviate the need for large technical teams.

Bibliography

N. Ancona and T. Poggio. Optical flow from 1D correlation: Application to a simple time-to-crash detector. In *Proceedings of Fourth International Conference on Computer Vision*, pp. 209-214, 1993.

S. Avadhanula, R. J. Wood, E. Steltz, J. Yan and R. S. Fearing. Lift force improvements for the micromechanical flying insect. In *IEEE Int. Conf. on Intelligent Robots and Systems*, pp. 1350-1356, 2003.

A. Bab-Hadiashar, D. Suter and R. Jarvis. Image-interpolation based optic flow technique. Technical Report MECSE-96-1, Monash University, Clayton, Australia, 1996.

E. Baird, M. V. Srinivasan, S. W. Zhang, R. Lamont and A. Cowling. Visual control of flight speed and height in the honeybee. In *From animals to animats 9: Proceedings of the Ninth International Conference on Simulation of Adaptive Behaviour, LNAI*, Vol. 4095. Springer-Verlag, 2006.

J. L. Barron, D. J. Fleet and S. S. Beauchemin. Performance of optical flow techniques. *International Journal of Computer Vision*, 12(1):43-77, 1994.

G. L. Barrows, C. Neely and K. T. Miller. Optic flow sensors for MAV navigation. In Thomas J. Mueller (Ed.), *Fixed and Flapping Wing Aerodynamics for Micro Air Vehicle Applications*, Volume 195 of *Progress in Astronautics and Aeronautics*, pp. 557-574. AIAA, 2001.

G. L. Barrows, J. S. Chahl and M. V. Srinivasan. Biomimetic visual sensing and flight control. In *Bristol Conference on UAV Systems*, 2002.

G. A. Bekey. *Autonomous Robots: From Biological Inspiration to Implementation and Control*. MIT Press, 2005.

A. Beyeler, C. Mattiussi, J.-C. Zufferey and D. Floreano. Vision-based altitude and pitch estimation for ultra-light indoor aircraft. In *IEEE International Conference on Robotics and Automation (ICRA'06)*, pp. 2836-2841, 2006.

A. Beyeler, J.-C. Zufferey and D. Floreano. Fast prototyping of optic-flow-based navigation strategies for autonomous indoor flight using simulation. In *IEEE International Conference on Robotics and Automation (ICRA'07)*, 2007.

A. Borst. How do flies land? from behavior to neuronal circuits. *BioScience*, 40(4):292-299, 1990.

A. Borst and S. Bahde. Spatio-temporal integration of motion. *Naturwissenschaften*, 75:265-267, 1988.

A. Borst and M. Egelhaaf. Principles of visual motion detection. *Trends in Neurosciences*, 12(8):297-306, 1989.

A. Borst, M. Egelhaaf and H. S. Seung. Two-dimensional motion perception in flies. *Neural Computation*, 5(6):856-868, 1993.

V. Braitenberg. *Vehicles – Experiments In Synthetic Psychology*. The MIT Press, Cambridge, MA, 1984.

R. A. Brooks. Achieving artificial intelligence through building robots. Technical Report A.I. Memo 899, Massachusetts Institute of Technology, Cambridge, MA, 1986.

R. A. Brooks. *Cambrian Intelligence*. The MIT Press, Cambridge, MA, 1999.

C. Cafforio and F. Rocca. Methods for measuring small displacements of television images. *IEEE Transactions on Information Theory*, 22:573- 579, 1976.

T. Camus. Calculating time-to-contact using real-time quantized optical flow. Technical Report 5609, National Institute Of Standards and Technology NISTIR, 1995.

J. S. Chahl and M. V. Srinivasan. Visual computation of egomotion using an image interpolation technique. *Biological Cybernetics*, 74:405-411, 1996.

J. S. Chahl, M. V. Srinivasan and H. Zhang. Landing strategies in honeybees and applications to uninhabited airborne vehicles. *The International Journal of Robotics Research*, 23(2):101-110, 2004.

W. P. Chan, F. Prete and M. H. Dickinson. Visual input to the efferent control system of a fly's "gyroscope". *Science*, 280(5361):289-292, 1998.

C. Chang and P. Gaudian (Eds), *Biomimetic Robotics*, Volume 30 of *Robotics and Autonomous Systems, Special Issue*. Elsevier, 2000.

R. F. Chapman. *The Insects: Structure and Function*. Cambridge University Press, 4th edition, 1998.

D. Cliff and G. F. Miller. Co-evolution of pursuit and evasion II: Simulation methods and results. In P. Maes, M. Mataric, J. A. Meyer, J. Pollack, H. Roitblat and S. Wilson (Eds), *From Animals to Animats IV: Proceedings of the Fourth International Conference on Simulation of Adaptive Behavior*. Cambridge, MA: MIT Press-Bradford Books, 1996.

D. Cliff, I. Harvey and P. Husbands. Artificial evolution of visual control systems for robots. In M. Srinivisan and S. Venkatesh (Eds), *From Living Eyes to Seeing Machines*, pp. 126-157. Oxford University Press, 1997.

T. S. Collett and M. F. Land. Visual control of flight behavior in the hoverfly, syritta pipiens. *Journal of Comparative Physiology*, 99:1-66, 1975.

D. Coombs, M. Herman, T. H. Hong and M. Nashman. Real-time obstacle avoidance using central flow divergence and peripheral flow. In *International Conference on Computer Vision*, pp. 276-283, 1995.

F. M. da Silva Metelo and L. R. Garcia Campos. Vision based control of an autonomous blimp. Technical report, 2003.

C. T. David. Compensation for height in the control of groundspeed by drosophila in a new, 'barber's pole' wind tunnel. *Journal of Comparative Physiology A*, 147:485-493, 1982.

T. Delbrück and C. A. Mead. Analog VLSI phototransduction by continuous-time, adaptive, logarithmic photoreceptor circuits. pp. 139-161. IEEE Computer Society Press, 1995.

M. H. Dickinson. Haltere-mediated equilibrium reflexes of the fruit fly, drosophila melanogaster. *Philosophical Transactions: Biological Sciences*, 354(1385):903-916, 1999.

M. H. Dickinson, F. O. Lehmann and S. P. Sane. Wing rotation and the aerodynamic basis of insect flight. *Science*, 284:1954-1960, 1999.

J. K. Douglass and N. J. Strausfeld. Visual motion-detection circuits in flies: parallel direction- and non-direction-sensitive pathways between the medulla and lobula plate. *Journal of Neuroscience*, 16(15):4551-4562, 1996.

R. O. Dror, D. C. O'Carroll and S. B. Laughlin. Accuracy of velocity estimation by Reichardt correlators. *Journal of Optical Society of America A*, 18:241-252, 2001.

A. P. Duchon and W. H. Warren. Robot navigation from a gibsonian viewpoint. In *Proceedings of IEEE Conference on Systems, Man and Cybernetics, San Antonio (TX)*, pp. 2272-2277, 1994.

A. P. Duchon, W. H. Warren and L. P. Kaelbling. Ecological robotics. *Adaptive Behavior*, 6:473-507, 1998.

R. Dudley. *The Biomechanics of Insect Flight: Form, Function, Evolution.* Princeton University Press, 2000.

J. Duparre, P. Schreiber, P. Dannberg, T. Scharf, P. Pelli, R. Volkel, H. P. Herzig and A. Brauer. Artificial compound eyes – different concepts and their application to ultra flat image acquisition sensors. In *Proceedings of SPIE, Vol. 5346, MOEMS and Miniaturized Systems IV*, 2004.

M. Egelhaaf and A. Borst. Transient and steady-state response properties of movement detectors. *Journal of Optical Society of America A*, 6(1):116-127, 1989.

M. Egelhaaf and A. Borst. A look into the cockpit of the fly: Visual orientation, algorithms, and identified neurons. *The Journal of Neuroscience*, 13 (11):4563-4574, 1993a.

M. Egelhaaf and A. Borst. Motion computation and visual orientation in flies. *Comparative Biochemistry and Physiology A*, 104(4):659-673, 1993b.

M. Egelhaaf and R. Kern. Vision in flying insects. *Current Opinion in Neurobiology*, 12(6):699-706, 2002.

M. Egelhaaf, R. Kern, H. G. Krapp, J. Kretzberg, R. Kurtz and A. K. Warzechna. Neural encoding of behaviourally relevant visual-motion information in the fly. *Trends in Neurosciences*, 25(2):96-102, 2002.

S. M. Ettinger, M. C. Nechyba, P. G. Ifju and M. Waszak. Vision-guided flight stability and control for micro air vehicles. *Advanced Robotics*, 17 (3):617-640, 2003.

H. R. Everett. *Sensors for Mobile Robots: Theory and Application.* AK Peters, Ltd. Natick, MA, USA, 1995.

R. S. Fearing, K. H. Chiang, M. H. Dickinson, D. L. Pick, M. Sitti and J. Yan. Wing transmission for a micromechanical flying insect. In *Proceeding of the IEEE International Conference on Robotics and Automation*, pp. 1509-1516, 2000.

C. Fennema and W. B. Thompson. Velocity determination in scenes containing several moving objects. *Computer Graphics and Image Processing*, 9: 301-315, 1979.

C. Fermüller and Y. Aloimonos. Primates, bees, and UGV's (unmanned ground vehicles) in motion. In M. Srinivisan and S. Venkatesh (Eds), *From Living Eyes to Seeing Machines*, pp. 199-225. Oxford University Press, 1997.

A. Fernandez Perez de Talens and C. T. Ferretti. *Landing and Optomotor Responses of the Fly Musca*, pp. 490-501. Clarendon Press, Oxford, 1975.

D. Floreano and C. Mattiussi. Evolution of spiking neural controllers for autonomous vision-based robots. In T. Gomi (Ed.), *Evolutionary Robotics IV*, pp. 38-61. Springer-Verlag, 2001.

D. Floreano, N. Schoeni, G. Caprari and J. Blynel. Evolutionary bits'n'spikes. In R. K. Standish, M. A. Beadau and H. A. Abbass (Eds), *Artificial Life VIII: Proceedings of the Eight International Conference on Artificial Life*. MIT Press, 2002.

D. Floreano, J.-C. Zufferey and C. Mattiussi. Evolving spiking neurons from wheels to wings. In K. Murase and T. Asakura (Eds), *Dynamic Systems Approach for Embodiment and Sociality*, Volume 6 of *Advanced Knowledge International, International Series on Advanced Intelligence*, pp. 65-70, 2003.

D. Floreano, T. Kato, D. Marocco and E. Sauser. Coevolution of active vision and feature selection. *Biological Cybernetics*, 90(3):218-228, 2004.

D. Floreano, J.-C. Zufferey and J. D. Nicoud. From wheels to wings with evolutionary spiking circuits. *Artificial Life*, 11(1-2):121-138, 2005.

D. Floreano, M. V. Srinivasan, C. P. Ellington and J.-C. Zufferey (Eds), *Proceedings of the International Symposium on Flying Insects and Robots, Monte Verità, Switzerland*, 2007.

N. Franceschini. Sampling of the visual environment by the compound eye of the fly: Fundamentals and applications. In A. W. Snyder and R. Menzel (Eds), *Photoreceptor Optics*, pp. 98-125. Springer, Berlin, 1975.

N. Franceschini. Visual guidance based on optic flow: a biorobotic approach. *Journal of Physiology – Paris*, 98:281-292, 2004.

N. Franceschini, A. Riehle and A. Le Nestour. Directionaly selective motion detection by insect neurons. In D. G. Stavenga and R. C. Hardie (Eds), *Facets of Vision*, pp. 360-390. Springer-Verlag, 1989.

N. Franceschini, J. M. Pichon and C. Blanes. From insect vision to robot vision. *Philosophical Transactions of the Royal Society B*, 337:283-294, 1992.

N. Franceschini, F. Ruffier and J. Serres. A bio-inpired flying robot sheds light on insect piloting abilities. *Current Biology*, 17:1-7, 2007.

M. O. Franz and J. S. Chahl. Insect-inspired estimation of self-motion. In *Proceedings of the 2nd International Workshop on Biologically Motivated Computer Vision, LNCS*, pp. 171-180. Springer-Verlag, 2002.

M. O. Franz and H. G. Krapp. Wide-field, motion-sensitive neurons and matched filters for optic flow fields. *Biological Cybernetics*, 83:185-197, 2000.

M. O. Franz and H. A. Mallot. Biomimetic robot navigation. *Robotics and Autonomous Systems*, 30:133-153, 2000.

S. Fry, R. Sayaman and M. H. Dickinson. The aerodynamics of free-flight maneuvers in drosophila. *Science*, 300:495-498, 2003.

J. J. Gibson. *The Perception of the Visual World*. Houghton Mifflin, Boston, 1950.

J. J. Gibson. *The Ecological Approach to Visual Perception*. Houghton Mifflin, Boston, 1979.

D. E. Goldberg. *Genetic Algorithms in Search, Optimization and Machine Learning*. Addison-Wesley, Reading, MA, 1989.

J. M. Grasmeyer and M. T. Keennon. Development of the black widow micro air vehicle. In Thomas J. Mueller (Ed.), *Fixed and Flapping Wing Aerodynamics for Micro Air Vehicle Applications*, Volume 195 of *Progress in Astronautics and Aeronautics*, pp. 519-535. AIAA, 2001.

W. E. Green, P. Y. Oh and G. L. Barrows. Flying insect inspired vision for autonomous aerial robot maneuvers in near-earth environments. In *Proceeding of the IEEE International Conference on Robotics and Automation*, Vol. 3, pp. 2347-2352, 2004.

S. Griffiths, J. Saunders, A. Curtis, T. McLain and R. Beard. *Obstacle and Terrain Avoidance for Miniature Aerial Vehicles*, Volume 33 of *Intelligent Systems, Control and Automation: Science and Engineering*, Chapter I.7, pp. 213-244. Springer, 2007.

K. G. Götz. The optomotor equilibrium of the drosophila navigation system. *Journal of Comparative Physiology*, 99:187-210, 1975.

J. Haag, M. Egelhaaf and A. Borst. Dendritic integration of motion information in visual interneurons of the blowfly. *Neuroscience Letters*, 140: 173-176, 1992.

R. R. Harrison and C. Koch. A robust analog VLSI motion sensor based on the visual system of the fly. *Autonomous Robots*, 7:211-224, 1999.

R. R. Harrison. *An Analog VLSI Motion Sensor Based on the Fly Visual System*. PhD thesis, 2000.

R. R. Harrison. An algorithm for visual collision detection in real-world scenes. Technical report, 2003.

I. Harvey, P. Husbands and D. Cliff. Seeing the light: Artificial evolution, real vision. In D. Cliff, P. Husbands, J. Meyer and S. Wilson (Eds), *From Animals to Animats III*, pp. 392-401. The MIT Press, 1994.

B. Hassenstein and W. Reichardt. Systemtheoretische analyse der zeit-, reihenfolgen- und vorzeichenauswertung bei der bewe-gungsperzeption des rüsselkäfers chlorophanus. *Zeitschrift für Naturforschung*, 11b:513-524, 1956.

K. Hausen. Motion sensitive interneurons in the optomotor system of the fly. II. the horizontal cells: Receptive field organization and response characteristics. *Biological Cybernetics*, 46:67-79, 1982.

K. Hausen and M. Egelhaaf. Neural mechanisms of visual course control in insects. In D. G. Stavenga and R. C. Hardie (Eds), *Facets of Vision*. Springer-Verlag, 1989.

R. Hengstenberg. Gaze control in the blowfly calliphora: A multisensory two-stage integration process. *The Neuroscience*, 3:19-29, 1991.

B. K. Horn. *Robot Vision*. MIT Press, 1986.

B. K. Horn and P. Schunck. Determining optical flow. *Artificial Intelligence*, 17:185-203, 1981.

A. Horridge. Insects which turn and look. *Endeavour*, 1:7-17, 1977.

A. Horridge. Pattern and 3D vision of insects. In Y. Aloimonos (Ed.), *Visual Navigation*, pp. 26-59. Lawrence Erlbaum Associates, Mahwah, New Jersey, 1997.

S. Hrabar, G. S. Sukhatme, P. Corke, K. Usher and J. Roberts. Combined optic-flow and stereo-based navigation of urban canyons for UAV. In *IEEE International Conference on Intelligent Robots and Systems*, pp. 3309-3316. IEEE, 2005.

S. A. Huber. *Studies of the Visual Orientation Behavior in Flies Using the Artificial Life Approach*. PhD thesis, Eberhard-Karls-Universität zu Tübingen, 1997.

S. A. Huber, H. A. Mallot and H. H. Bülthoff. Evolution of the sensorimotor control in an autonomous agent. In *Proceedings of the Fourth International Conference on Simulation of Adaptive Behavior*, pp. 449-457. MIT Press, 1996.

F. Iida. Goal-directed navigation of an autonomous flying robot using biologically inspired cheap vision. In *Proceedings of the 32nd International Symposium on Robotics*, 2001.

F. Iida. Biologically inspired visual odometer for navigation of a flying robot. *Robotics and Autonomous Systems*, 44:201-208, 2003.

F. Iida and D. Lambrinos. Navigation in an autonomous flying robot by using a biologically inspired visual odometer. In *Sensor Fusion and Decentralized Control in Robotic System III, Photonics East, Proceeding of SPIE*, Vol. 4196, pp. 86-97, 2000.

K. D. Jones, C. J. Bradshaw, J. Papadopoulos and M. F. Platzer. Improved performance and control of flapping-wing propelled micro air vehicles. In *Proceedings of the 42nd Aerospace Sciences Meeting and Exhibit, Reno NV*, 2004.

K. Karmeier, M. Egelhaaf and H. G. Krapp. Early visual experience and receptive field organization of the optic flow processing interneurons in the fly motion pathway. *Visual Neuroscience*, 18:1-8, 2001.

J. S. Kennedy. The migration of the desert locust. *Philosophical transactions of the Royal Society of London B*, 235:163-290, 1951.

J. J. Koenderink and A. J. van Doorn. Facts on optic flow. *Biological Cybernetics*, 56:247-254, 1987.

J. Kramer, R. Sarpeshkar and C. Koch. An analog VLSI velocity sensor. In *Proceedings of IEEE International Symposium on Circuits and Systems*, pp. 413-416, 1995.

H. G. Krapp. Neuronal matched filters for optic flow processing in flying insects. In M. Lappe (Ed.), *Neuronal Processing of Optic Flow*, pp. 93-120. San Diego: Accademic Press, 2000.

H. G. Krapp and R. Hengstenberg. Estimation of self-motion by optic flow processing in single visual interneurons. *Nature*, 384:463-466, 1996.

H. G. Krapp, B. Hengstenberg and R. Hengstenberg. Dendritic structure and receptive-field organization of optic flow processing interneurons in the fly. *Journal of Neurophysiology*, 79:1902-1917, 1998.

I. Kroo and P. Kunz. Mesoscale flight and miniature rotorcraft development. In Thomas J. Mueller (Ed.), *Fixed and Flapping Wing Aerodynamics for Micro Air Vehicle Applications*, Volume 195 of *Progress in Astronautics and Aeronautics*, pp. 503-517. AIAA, 2001.

D. Lambrinos, R. Möller, T. Labhart, R. Pfeifer and R. Wehner. A mobile robot employing insect strategies for navigation. *Robotics and Autonomous Systems*, 30:39-64, 2000.

M. F. Land. Visual acuity in insects. *Annual Review of Entomology*, 42:147-177, 1997.

D. N. Lee. A theory of visual control of braking based on information about time-to-collision. *Perception*, 5:437-459, 1976.

D. N. Lee and P. E. Reddish. Plummeting gannets: a paradigm of ecological optics. *Nature*, 293:293-294, 1981.

F. O. Lehmann. Aerial locomotion in flies and robots: kinematic control and aerodynamics of oscillating wings. *Arthropod Structure and Development*, 33:331-345, 2004.

D. Lentink. Novel micro aircraft inspired by insect flight. In D. Floreano, M. V. Srinivasan, C. P. Ellington and J.-C. Zufferey (Eds), *Proceedings* of the International Symposium on Flying Insects and Robots, Monte Verità, Switzerland, pp. 67-68, 2007.

M. A. Lewis. Visual navigation in a robot using zig-zag behavior. In *Neural Information Processing Systems 10*. MIT Press, 1998.

L. Lichtensteiger and P. Eggenberger. Evolving the morphology of a compound eye on a robot. In *Proceedings of the Third European Workshop on Advanced Mobile Robots (Eurobot '99)*, pp. 127-134, 1999.

J. P. Lindemann, R. Kern, C. Michaelis, J.-A. Meyer, J. H. van Hateren and M. Egelhaaf. Flimax, a novel stimulus device for panoramic and highspeed presentation of behaviourally generated optic flow. *Vision Research*, 43:779-791, 2003.

S.-C. Liu, J. Kramer, G. Indiveri, T. Delbrück and R. Douglas. *Analog VLSI: Circuits and Principles*. MIT Press, Cambridge, MA, 2003.

B. Lucas and T. Kanade. An iterative image registration technique with an application to stereo vision. In *Proceedings of the Seventh International Joint Conference on Artificial Intelligence, Vancouver*, pp. 674-679, 1981.

H. A. Mallot. *Computational Vision: Information Processing in Perception and Visual Behavior*. The MIT Press, 2000.

D. Marocco and D. Floreano. Active vision and feature selection in evolutionary behavioral systems. In J. Hallam, D. Floreano, G. Hayes and J. Meyer (Eds), *From Animals to Animats 7: Proceedings of the Seventh International Conference on Simulation of Adaptive Behavior*, pp. 247-255, Cambridge, MA, 2002. MIT Press-Bradford Books.

D. Marr. *Vision: A Computational Investigation into the Human Representation and Processing of Visual Information*. W. H. Freeman and Company, New York, 1982.

J. H. McMasters and M. L. Henderson. Low speed single element airfoil synthesis. *Technical Soaring*, 2:1-21, 1980.

C. Melhuish and J. Welsby. Gradient ascent with a group of minimalist real robots: Implementing secondary swarming. In *Proceedings of the IEEE International Conference on Systems, Man and Cybernetics*, 2002.

O. Michel. Webots: Professional mobile robot simulation. *International Journal of Advanced Robotic Systems*, 1(1):39-42, 2004.

R. Moeckel and S.-C. Liu. Motion detection circuits for a time-to-travel algorithm. In *IEEE International Symposium on Circuits and Systems (ISCAS)*, 2007.

F. Mondada, E. Franzi and P. Ienne.   Mobile robot miniaturization: a tool for investigation in control algorithms.   In T. Yoshikawa and F. Miyazaki (Eds), *Proceedings of the Third International Symposium on Experimental Robotics, Kyoto, Japan*, pp. 501-513. Springer-Verlag, 1993.

T. J. Mueller. *Fixed and Flapping Wing Aerodynamics for Micro Air Vehicle Applications*, Volume 195 of *Progress in Astronautics and Aeronautics*. AIAA, 2001.

T. J. Mueller and J. D. DeLaurier.   An overview of micro air vehicle.   In T. J. Mueller (Ed.), *Fixed and Flapping Wing Aerodynamics for Micro Air Vehicle Applications*, Progress in Astronautics and Aeronautics, pp. 1-10. AIAA, 2001.

F. Mura and N. Franceschini. Visual control of altitude and speed in a flying agent. In *From Animals to Animats III*, pp. 91-99. MIT Press, 1994.

L. Muratet, S. Doncieux, Y. Brière and J. A. Meyer.   A contribution to vision-based autonomous helicopter flight in urban environments. *Robotics and Autonomous Systems*, 50(4):195-209, 2005.

H. H. Nagel. On change detection and displacement vector estimation in image sequences. *Pattern Recognition Letters*, 1:55-59, 1982.

M. G. Nagle and M. V. Srinivasan. Structure from motion: determining the range and orientation of surfaces by image interpolation. *Journal of the Optical Society of America A*, 13(1):5-34, 1996.

G. Nalbach.   The halteres of the blowfly calliphora. I. Kinematics and dynamics. *Journal of Comparative Physiology A*, 173(3):293-300, 1993.

G. Nalbach. Extremely non-orthogonal axes in a sense organ for rotation: Behavioural analysis of the dipteran haltere system. *Neuroscience*, 61:49-163, 1994.

G. Nalbach and R. Hengstenberg. The halteres of the blowfly calliphora. II. Three-dimensional organization of compensatory reactions to real and simulated rotations. *Journal of Comparative Physiology A*, 175(6):695-708, 1994.

R. C. Nelson and Y. Aloimonos. Finding motion parameters from spherical flow fields (or the advantages of having eyes in the back of your head). *Biological Cybernetics*, 58:261-273, 1988.

R. C. Nelson and Y. Aloimonos. Obstacle avoidance using flow field divergence. *IEEE Transactions on Pattern Analysis and Machine Intelligence*, 11 (10):1102-1106, 1989.

T. Netter and N. Franceschini. A robotic aircraft that follows terrain using a neuromorphic eye. In *Proceedings of the IEEE/RSJ International Conference on Intelligent Robots and Systems*, 2002.

T. R. Neumann. Modeling instect compound eyes: Space-variant spherical vision. In H. H. Bülthoff, S.-W. Lee, T. Poggio and C. Wallraven (Eds), *Proceedings of the 2nd International Workshop on Biologically Motivated Computer Vision*. Springer-Verlag, 2002.

T. R. Neumann. *Biometric Spherical Vision*. Logos Berlin, 2003.

T. R. Neumann and H. H. Bülthoff. Insect inspired visual control of translatory flight. In J. Kelemen and P. Sosík (Eds), *Advances in Artificial Life : 6th European Conference, ECAL 2001, Prague, Czech Republic*, Volume 2159 of *Lecture Notes in Computer Science*. Springer-Verlag, 2001.

T. R. Neumann and H. H. Bülthoff. Behavior-oriented vision for biomimetic flight control. In *Proceedings of the EPSRC/BBSRC International Workshop on Biologically Inspired Robotics*, pp. 196-203, 2002.

T. R. Neumann, S. A. Huber and H. H. Bülthoff. Minimalistic approach to 3d obstacle avoidance behavior from simulated evolution. In *Proceedings of the 7th International Conference on Artificial Neural Networks (ICANN)*, Volume 1327 of *Lecture Notes in Computer Science*, pp. 715-720. Springer-Verlag, 1997.

J.-D. Nicoud and J.-C. Zufferey. Toward indoor flying robots. *IEEE/RSJ International Conference on Robots and Systems*, pp. 787-792, 2002.

S. Nolfi and D. Floreano. *Evolutionary Robotics. The Biology, Intelligence and Technology of Self-organizing Machines*. MIT Press, Boston, MA, 2000.

J. M. Pichon, C. Blanes and N. Franceschini. Visual guidance of a mobile robot equipped with a network of self-motion sensors. *Proceedings of SPIE: Mobile Robots IV*, 1195:44-53, 1990.

C. Planta, J Conradt, A. Jencik and P. Verschure. A neural model of the fly visual system applied to navigational tasks. In *Proceedings of the International Conference on Artificial Neural Networks (ICANN)*, 2002.

T. Poggio, A. Verri and V. Torre. Green theorems and qualitative properties of the optical flow. Technical Report A.I. Memo 1289, Massachusetts Institute of Technology, Cambridge, MA, 1991.

T. N. Pornsin-Sirirak, Y.-Ch. Tai and Ch.-M. Ho. Microbat: A palm-sized electrically powered ornithopter. In *Proceedings of NASA/JPL Workshop on Biomorphic Robotics*, 2001.

W. Reichardt. Autocorrelation, a principle for the evaluation of sensory information by the central nervous system. In W. A. Rosenblith (Ed.), *Sensory Communication*, pp. 303-317. Wiley, New York, 1961.

W. Reichardt. Movement perception in insects. In W. Reichardt (Ed.), *Processing of Optical Data by Organisms and by Machines*, pp. 465-493. New York: Academic Press, 1969.

M. B. Reiser and M. H. Dickinson. A test bed for insect-inspired robotic control. *Philosophical Transactions: Mathematical, Physical & Engineering Sciences*, 361(1811):2267-2285, 2003.

F. C. Rind. Motion detectors in the locust visual system: From biology to robot sensors. *Microscopy Research and Technique*, 56(4):256-269, 2002.

F. Ruffier. *Pilote automatique biomimetique*. PhD thesis, 2004.

F. Ruffier and N. Franceschini. Octave, a bioinspired visuo-moto control system for the guidance of micro-air-vehicles. In A. Rodriguez-Vazquez, D. Abott and R. Carmona (Eds), *Proceedings of SPIE Conference on Bioengineered and Bioinspired Systems*, pp. 1-12, 2003.

F. Ruffier and N. Franceschini. Visually guided micro-aerial vehicle: automatic take off, terrain following, landing and wind reaction. In *Proceedings of the IEEE International Conference on Robotics and Automation*, pp. 2339-2346. T. J. Tarn and T. Fukuda and K. Valavanis, April 2004.

S. P. Sane, A. Dieudonné, M. A. Willis and T. L. Daniel. Antennal mechanosensors mediate flight control in moths. *Science*, 315:863-866, 2007.

J. Santos-Victor, G. Sandini, F. Curotto and S. Garibaldi. Divergent stereo for robot navigation: A step forward to a robotic bee. *International Journal of Computer Vision*, 14:159-177, 1995.

L. Schenato, W. C. Wu and S. S. Sastry. Attitude control for a micromechanical flying insect via sensor output feedback. *IEEE Journal of Robotics and Automation*, 20(1):93-106, 2004.

S. Scherer, S. Singh, L. Chamberlain and S. Saripalli. Flying fast and low among obstacles. In *Proceedings of the 2007 IEEE Conference on Robotics and Automation*, pp. 2023-2029, 2007.

C. Schilstra and J. H. van Hateren. Blowfly flight and optic flow. I. thorax kinematics and flight dynamics. *Journal of Experimental Biology*, 202: 1481-1490, 1999.

H. Schuppe and R. Hengstenberg. Optical properties of the ocelli of calliphora erythrocephala and their role in the dorsal light response. *Journal of Comparative Physiology A*, 173:143-149, 1993.

A. Sherman and M. H. Dickinson. Summation of visual and mechanosensory feedback in drosophila flight control. *Journal of Experimental Biology*, 207:133-142, 2004.

R. Siegwart and I. Nourbakhsh. *Introduction to Autonomous Mobile Robotics*. MIT Press, 2004.

S. Single, J. Haag and A. Borst. Dendritic computation of direction selectivity and gain control in visual interneurons. *Journal of Neuroscience*, 17: 6023-6030, 1997.

E. Sobel. The locust's use of motion parallax to measure distance. *Journal of Comparative Physiology A*, 167:579-588, 1990.

P. Sobey. Active navigation with a monocular robot. *Biological Cybernetics*, 71:433-440, 1994.

M. V. Srinivasan. An image-interpolation technique for the computation of optic flow and egomotion. *Biological Cybernetics*, 71:401-416, 1994.

M. V. Srinivasan. Visual navigation: The eyes know where their owner is going. In *Motion Vision: Computational, Neural and Ecological Constraints*, pp. 180-186. Springer, 2000.

M. V. Srinivasan and H. Zhang. Visual navigation in flying insects. *International Review of Neurobiology*, 44:67-92, 2000.

M. V. Srinivasan, M. Lehrer, W. H. Kirchner and S. W. Zhang. Range perception through apparent image speed in freely-flying honeybees. *Visual Neuroscience*, 6:519-535, 1991.

M. V. Srinivasan, S. W. Zhang, M. Lehrer and T. S. Collett. Honeybee navigation en route to the goal: Visual flight control and odometry. *The Journal of Experimental Biology*, 199:237-244, 1996.

M. V. Srinivasan, J. S. Chahl, M. G. Nagle and S. W. Zhang. Embodying natural vision into machines. In M.V. Srinivasan and S. Venkatesh (Eds), *From Living Eyes to Seeing Machines*, pp. 249-265. 1997.

M. V. Srinivasan, J. S. Chahl, K. Weber, S. Venkatesh and H. Zhang. Robot navigation inspired by principles of insect vision. In A. Zelinsky (Ed.), *Field and Service Robotics*, pp. 12-16. Springer-Verlag, 1998.

M. V. Srinivasan, M. Poteser and K. Kral. Motion detection in insect orientation and navigation. *Vision Research*, 39(16):2749-2766, 1999.

M. V. Srinivasan, S. W. Zhang, J. S. Chahl, E. Barth and S. Venkatesh. How honeybees make grazing landings on flat surfaces. *Biological Cybernetics*, 83:171-183, 2000.

N. J. Strausfeld. Beneath the compound eye. Neuroanatomical and physiological correlates in the study of insect vision. D. G. Stavenga and R. C. Hardie (Eds), *Facets of Vision*, Springer-Verlag, 1989.

L. F. Tammero and M. H. Dickinson. The influence of visual landscape on the free flight behavior of the fruit fly drosophila melanogaster. *The Journal of Experimental Biology*, 205:327-343, 2002a.

L. F. Tammero and M. H. Dickinson. Collision-avoidance and landing responses are mediated by separate pathways in the fruit fly. *The Journal of Experimental Biology*, 205:2785-2798, 2002b.

S. Thakoor, J. M. Morookian, J. S. Chahl, B. Hine and S. Zornetzer. Bees: Exploring mars with bioinspired technologies. *Computer*, 37(9):38-47, 2004.

S. Thrun, W. Burgard and D. Fox. *Probabilistic Robotics*. MIT Press, 2005.

J. Urzelai and D. Floreano. Evolution of adaptive synapses: robots with fast adaptive behavior in new environments. *Evolutionary Computation*, 9(4): 495-524, 2001. doi: 10.1162/10636560152642887.

S. van der Zwaan, A. Bernardino and J. Santos-Victor. Visual station keeping for floating robots in unstructured environments. *Robotics and Autonomous Systems*, 39:145-155, 2002.

J. H. van Hateren and C. Schilstra. Blowfly flight and optic flow. II. head movements during flight. *Journal of Experimental Biology*, 202:1491-1500, 1999.

A. Verri, M. Straforini and V. Torre. Computational aspects of motion perception in natural and artificial vision systems. *Philosophical Transactions of the Royal Society B*, 337:429-443, 1992.

H. Wagner. Flow-field variables trigger landing in flies. *Nature*, 297:147-148, 1982.

H. Wagner. Flight performance and visual control of flight of the free-flying housefly (Musca domestica L.). I. organization of the flight motor. *Philosophical Transactions of the Royal Society B*, 312:527-551, 1986.

B. Webb and T. R. Consi. *Biorobotics: Methods and Applications*. MIT Press, 2001.

B. Webb, R. R. Harrison and M. A. Willis. Sensorimotor control of navigation in arthropod and artificial systems. *Arthropod Structure and Development*, 33:301-329, 2004.

K. Weber, S. Venkatesh and M. V. Srinivasan. Insect inspired behaviours for the autonomous control of mobile robots. In M. V. Srinivasan and S. Venkatesh (Eds), *From Living Eyes to Seeing Machines*, pp. 226-248. Oxford University Press, 1997.

R. Wehner. Matched filters – neural models of the external world. *Journal of Comparative Physiology A*, 161:511-531, 1987.

T. C. Whiteside and G. D. Samuel. Blur zone. *Nature*, 225:94-95, 1970.

W. C. Wu, L. Schenato, R. J. Wood and R. S. Fearing. Biomimetic sensor suite for flight control of a micromechanical flight insect: Design and experimental results. In *Proceeding of the IEEE International Conference on Robotics and Automation, Taipei, Taiwan*, pp. 1146-1151, 2003.

H. Zhang and J. P. Ostrowski. Visual servoing with dynamics: Control of an unmanned blimp. Technical report, 1998.

J.-C. Zufferey, C. Halter and J.-D. Nicoud. Avion d'intérieur : une plate-forme de recherche pour la robotique bio-inspirée. *Modell Flugsport* (Swiss Magazine), 2001.

J.-C. Zufferey, D. Floreano, M. van Leeuwen and T. Merenda. Evolving vision-based flying robots. In Bülthoff, Lee, Poggio and Wallraven (Eds), *Biologically Motivated Computer Vision: Second International Workshop, BMCV 2002, Tžbingen, Germany*, Volume 2525 of *Lecture Notes in Computer Science*, pp. 592-600. Springer-Verlag, 2002.

J.-C. Zufferey, A. Beyeler and D. Floreano. Vision-based navigation from wheels to wings. In *IEEE/RSJ International Conference on Intelligent Robots and Systems*, Vol. 3, pp. 2968-2973, 2003.

J.-C. Zufferey, A. Klaptocz, A. Beyeler, J.-D. Nicoud and D. Floreano. A 10-gram vision-based flying robot. *Advanced Robotics, Journal of the Robotics Society of Japan*, 21(14):1671-1684, 2007.

J.-C. Zufferey, A. Guanella, A. Beyeler and D. Floreano. Flying over the reality gap: From simulated to real indoor airships. *Autonomous Robots*, 21(3):243-254, 2006.

J.-C. Zufferey and D. Floreano. Fly-inspired visual steering of an ultralight indoor aircraft. *IEEE Transactions on Robotics*, 22(1):137-146, 2006.

Index